T0275920

INTEGRATION FOR CALCULUS, ANALYSIS, AND DIFFERENTIAL EQUATIONS Techniques, Examples, and Exercises

INTEGRATION FOR CALCULUS, ANALYSIS, AND DIFFERENTIAL EQUATIONS Techniques, Examples, and Exercises

Marat V Markin

California State University, Fresno, USA

 **World Scientific**

NEW JERSEY · LONDON · SINGAPORE · BEIJING · SHANGHAI · HONG KONG · TAIPEI · CHENNAI · TOKYO

Published by

World Scientific Publishing Co. Pte. Ltd.

5 Toh Tuck Link, Singapore 596224

USA office: 27 Warren Street, Suite 401-402, Hackensack, NJ 07601

UK office: 57 Shelton Street, Covent Garden, London WC2H 9HE

Library of Congress Cataloging-in-Publication Data
Names: Markin, Marat V., author.
Title: Integration for calculus, analysis, and differential equations : techniques, examples,
    and exercises / by Marat V. Markin (California State University, Fresno, USA).
Description: New Jersey : World Scientific, 2018. | Includes bibliographical references and index.
Identifiers: LCCN 2018026833 | ISBN 9789813272033 (hardcover : alk. paper) |
    ISBN 9789813275157 (pbk : alk. paper)
Subjects: LCSH: Calculus--Textbooks. | Mathematical analysis--Textbooks. |
    Differential equations--Textbooks.
Classification: LCC QA303.2 .M368 2018 | DDC 515--dc23
LC record available at https://lccn.loc.gov/2018026833

British Library Cataloguing-in-Publication Data
A catalogue record for this book is available from the British Library.

For any available supplementary material, please visit
https://www.worldscientific.com/worldscibooks/10.1142/11035#t=suppl

Printed in Singapore

To my mother, Svetlana A. Markina, fondly.

Preface

> *The calculus is the story this*
> *world first told itself as it*
> *became the modern world.*

David Berlinski

Amply demonstrated by experience, *integral calculus* covered in *Calculus I* and, mostly, in *Calculus II* appears to represent a serious challenge for many students. Passing rates in these courses are often considered to be indicative for the future graduation rates.

The main purpose of this book is to assist calculus students to gain a better understanding and command of integration and its applications and, thus, improving their performance in *Calculus I* and *II* courses. Its writing stems out of my extensive experience of teaching calculus or its equivalent to diverse groups of students at the California State University, Fresno, Boston University, the University of North Carolina, Asheville, and the National University of Food Technologies, Kiev, Ukraine.

The usefulness of the book as a concise and, at the same time, rather comprehensive review of integration reaches beyond the scope of the foregoing courses to students in more advanced courses such as *Multivariable Calculus, Differential Equations*, and *Intermediate Analysis*, where the ability to effectively integrate is essential for their success, and also those, who prepare for integration competitions such as the *Fresno State Integration Bee*.

Keeping the reader constantly focused on the three principal epistemological questions: *What for? Why? How?*, the book is designated as a supplementary instructional tool treating the three kinds of integral: *indefinite, definite*, and *improper* and covering various aspects of integral calculus from abstract definitions and theorems (with complete proof whenever appropriate) through various integration techniques to applications. It contains 143 Examples, including 112 Problems with complete step-by-step solutions, the same problem occasionally solved in more than one way while encouraging the reader to find the most efficient integration path, 6 Exercises, 162 Practice Problems, and 30 Mixed Integration Problems "for dessert", where the reader is expected to independently choose and implement the best possible integration approach. The answers to all the 192 Problems are provided in the Answer Key. Three Appendices furnish a table of basic integrals, reduction formulas, and basic identities of algebra and trigonometry.

The book's writing was supported by a Fresno State College of Science and Mathematics Scholarly and Creative Activity Award 2015/16, for which I would like to express my cordial gratitude.

My utmost appreciation goes to Dr. Maria Nogin (Department of Mathematics, CSU, Fresno) for her numerous invaluable contributions into improving the manuscript and to Mr. Andres Zumba Quezada (CSU, Fresno), the winner of the *Fresno State Integration Bee* 2015 and 2017, for painstakingly reading the manuscript, solving every single problem in it, and providing helpful suggestions. I am also very grateful to Dr. Przemyslaw Kajetanowicz (Department of Mathematics, CSU, Fresno) for his kind assistance with the figures.

My sincere acknowledgments are also due to the following associates of *World Scientific Publishing Co. Pte. Ltd.*: Ms. Rochelle Kronzek for discerning a value in my manuscript and making the authors, in particular this one, her high priority; Ms. Lai Fun Kwong for her astounding efficiency and great editorial work, and to Mr. Rajesh Babu for his expert and very helpful technical assistance.

The book, my first one, is dedicated to my mother, Svetlana A. Markina, with affection and appreciation inexpressible with any words.

Marat V. Markin

Contents

Chapter 1

Indefinite and Definite Integrals

1.1. Antiderivatives and Indefinite Integral

1.1.1. *Definitions and Examples*

Definition 1.1 (Antiderivative).
Let f be a function defined on an interval I. A function F is called an *antiderivative* of $f(x)$ on I if

$$F'(x) = f(x) \text{ for all } x \text{ in } I.$$

Examples 1.1 (Antiderivatives).

1. The function $F(x) = 1$ is an *antiderivative* of $f(x) = 0$ on $(-\infty, \infty)$ as well as any function of the form $F(x) = C$, where C is an arbitrary real constant (written henceforth as $C \in \mathbb{R}$).
2. The function $F(x) = x$ is an *antiderivative* of $f(x) = 1$ on $(-\infty, \infty)$ as well as any function of the form $F(x) = x + C$, $C \in \mathbb{R}$.
3. The function $F(x) = \dfrac{x^2}{2}$ is an *antiderivative* of $f(x) = x$ on $(-\infty, \infty)$ as well as any function of the form $F(x) = \dfrac{x^2}{2} + C$, $C \in \mathbb{R}$.
4. Any function of the form $F(x) = e^x + C$, $C \in \mathbb{R}$, is an *antiderivative* of $f(x) = e^x$ on $(-\infty, \infty)$.

All the above examples have one thing in common: if F is an antiderivative of f on I, then so is any function of the form

$$F(x) + C, \ x \in I, \tag{1.1}$$

where C is an arbitrary real constant ($C \in \mathbb{R}$).
The natural question is: are there antiderivatives of f on I not included in this description? The answer is NO.

As follows from the *Mean Value Theorem* (see, e.g., [1, 6]), functions with the same derivative differ by a constant. Thus, if G is an arbitrary antiderivative of f on I, there is a $C \in \mathbb{R}$ such that

$$G(x) = F(x) + C, \ x \in I,$$

and hence, expression (1.1) describes *all possible antiderivatives* of f on I.

Definition 1.2 (Indefinite Integral).
Let a function f defined on an interval I have an antiderivative F on I. The *indefinite integral* (or the *general antiderivative*) of f on I is the expression

$$\int f(x)\,dx := F(x) + C, \ x \in I,$$

where C is an arbitrary real constant $(C \in \mathbb{R})$.

The integral notation $\int f(x)\,dx$, uses the *integral symbol* $\int$ and *differential symbol* dx. The function f is called the *integrand* and x the *integration variable*.

The process of finding an indefinite integral is called *integration* (or *antidifferentiation*).

Obviously, integration (antidifferentiation) is the process inverse to differentiation, i.e.,

$$\int f'(x)\,dx = f(x) + C \quad \text{and} \quad \frac{d}{dx} \int f(x)\,dx = f(x).$$

Thus, the following examples are readily obtained by reversing the standard table of basic derivatives with some natural minor adjustments, when required, and become a part of our *Table of Basic Integrals* (Appendix A).

Examples 1.2 (Basic Indefinite Integrals).

1. $\displaystyle\int 0\,dx = C$ on $(-\infty, \infty)$.

2. $\displaystyle\int x^n\,dx = \frac{x^{n+1}}{n+1} + C \ (n \neq -1)$

   on interval(s) depending on the exponent n.

   In particular, $\displaystyle\int 1\,dx = \int x^0\,dx = x + C, \ \int x\,dx = \frac{x^2}{2} + C$

   on $(-\infty, \infty)$,

$$\int \frac{1}{x^2}\, dx = \int x^{-2}\, dx = -x^{-1} + C$$

on each of the intervals $(-\infty, 0)$, $(0, \infty)$,

$$\int \sqrt{x}\, dx = \int x^{1/2}\, dx = \frac{x^{3/2}}{3/2} = \frac{2}{3}x^{3/2} + C \quad \text{on } [0, \infty),$$

3. $\int \frac{1}{x}\, dx = \ln |x| + C$ on each of the intervals $(-\infty, 0)$, $(0, \infty)$.

4. $\int a^x\, dx = \dfrac{a^x}{\ln a} + C$ $(a > 0, a \neq 1)$ on $(-\infty, \infty)$.

In particular, $\int e^x\, dx = e^x + C$ on $(-\infty, \infty)$.

5. $\int \sin x\, dx = -\cos x + C$ on $(-\infty, \infty)$.

6. $\int \cos x\, dx = \sin x + C$ on $(-\infty, \infty)$.

7. $\int \sec^2 x\, dx = \tan x + C$ on each of the intervals

$(-\pi/2 + n\pi, \pi/2 + n\pi)$, $n \in \mathbb{Z} := \{0, \pm 1, \pm 2, \dots\}$.

8. $\int \csc^2 x\, dx = -\cot x + C$ on each of the intervals $(n\pi, \pi + n\pi)$, $n \in \mathbb{Z}$.

9. $\int \sec x \tan x\, dx = \sec x + C$ on the same intervals as in 7.

10. $\int \csc x \cot x\, dx = -\csc x + C$ on the same intervals as in 8.

11. $\int \sinh x\, dx = \cosh x + C$ on $(-\infty, \infty)$.

12. $\int \cosh x\, dx = \sinh x + C$ on $(-\infty, \infty)$.

13. $\int \dfrac{1}{x^2 + 1}\, dx = \arctan x + C$ on $(-\infty, \infty)$.

14. $\int \dfrac{1}{\sqrt{1 - x^2}}\, dx = \arcsin x + C$ on $(-1, 1)$.

15. $\int \dfrac{1}{x\sqrt{x^2 - 1}}\, dx = \operatorname{arcsec} |x| + C$

on each of the intervals $(-\infty, -1)$, $(1, \infty)$

since, for $x > 1$,

$$\int \frac{1}{x\sqrt{x^2-1}}\,dx = \int \frac{1}{|x|\sqrt{x^2-1}}\,dx = \operatorname{arcsec}|x| + C$$

and, for $x < -1$,

$$\int \frac{1}{x\sqrt{x^2-1}}\,dx = -\int \frac{1}{|x|\sqrt{x^2-1}}\,dx$$
$$= -\operatorname{arcsec} x + B = \operatorname{arcsec}|x| - \pi + B = \operatorname{arcsec}|x| + C \quad (C := -\pi+B).$$

1.1.2. *Validation of Indefinite Integrals*

To validate the equality

$$\int f(x)\,dx = F(x) + C \text{ on } I$$

is to show that

$$F'(x) = f(x) \text{ on } I,$$

i.e., integration is validated via differentiation.

Examples 1.3 (Appendix A, integrals 11, 13, and 20).

1. $\int \tan x\,dx = \ln|\sec x| + C = -\ln|\cos x| + C$ on each of the intervals $(-\pi/2 + n\pi, \pi/2 + n\pi)$, $n \in \mathbb{Z}$, since, on each of these intervals,

$$[\ln|\sec x|]' = \frac{\sec' x}{\sec x} = \frac{\sec x \tan x}{\sec x} = \tan x$$

and

$$\ln|\sec x| = \ln\left[|\cos x|\right]^{-1} = -\ln|\cos x|.$$

2. $\int \sec x\,dx = \ln|\sec x + \tan x| + C$ on the same intervals as in the prior example since, on each of these intervals,

$$[\ln|\sec x + \tan x|]' = \frac{(\sec x + \tan x)'}{\sec x + \tan x} = \frac{\sec x \tan x + \sec^2 x}{\sec x + \tan x} = \sec x.$$

3. For $a > 0$,

$$\int \frac{1}{\sqrt{x^2 \pm a^2}}\,dx = \ln|x + \sqrt{x^2 \pm a^2}| + C$$

on $(-\infty, \infty)$ for "+" and on each of the intervals $(-\infty, -a)$, (a, ∞) for "−" since, on each of these intervals,

$$\frac{d}{dx}\ln|x+\sqrt{x^2\pm a^2}| = \frac{\left[x+\sqrt{x^2\pm a^2}\right]'}{x+\sqrt{x^2\pm a^2}} = \frac{1+\dfrac{\left[x^2\pm a^2\right]'}{2\sqrt{x^2\pm a^2}}}{x+\sqrt{x^2\pm a^2}}$$

$$= \frac{1+\dfrac{2x}{2\sqrt{x^2\pm a^2}}}{x+\sqrt{x^2\pm a^2}} = \frac{\dfrac{\sqrt{x^2\pm a^2}+x}{\sqrt{x^2\pm a^2}}}{x+\sqrt{x^2\pm a^2}} = \frac{1}{\sqrt{x^2\pm a^2}}.$$

1.1.3. Which Functions Are Integrable?

As follows from the *Fundamental Theorem of Calculus* (see Sec. 1.2.6), each function f *continuous* on an interval I has an antiderivative F on I (i.e., is *integrable* on I). For a function f having discontinuities on an interval I, this need not be true.

Exercise 1.1 (Sign Function).
Show that the *sign function*

$$\text{sgn}(x) := \begin{cases} -1 & \text{for } x < 0 \\ 0 & \text{for } x = 0 \\ 1 & \text{for } x > 0 \end{cases}$$

has antiderivatives on $(-\infty, 0)$, $(0, \infty)$, but has no antiderivative on any open interval I containing 0.

1.1.4. Properties of Indefinite Integral (Integration Rules)

Integration is governed by the following rules readily proved via differentiation.

Theorem 1.1 (Properties of Indefinite Integral (Integration Rules)).
If $\int f(x)\,dx = F(x) + C$ and $\int g(x)\,dx = G(x) + C$ on an interval I, then, on I,

1. for any $c \in \mathbb{R}$, $\quad \int cf(x)\,dx = cF(x) + C = c\int f(x)\,dx \qquad$ (*Constant Factor Rule*)

2. $\int [f(x) \pm g(x)]\,dx = F(x) \pm G(x) + C = \int f(x)\,dx \pm \int g(x)\,dx$ (*Sum/Difference Rule*)

Remark 1.1. For integration, unlike differentiation, the are NO *Product/ Quotient Rules.*

Exercise 1.2 (No Product/Quotient Rules for Integration).
Give an example of functions f and g having antiderivatives F and G on an interval I, respectively, for which

$$\int f(x)g(x)\,dx \neq F(x)G(x) + C \quad \text{and} \quad \int \frac{f(x)}{g(x)}\,dx \neq \frac{F(x)}{G(x)} + C \quad \text{on } I.$$

Example 1.4 (Using Integration Rules).

$$\int \left(\frac{4}{x^2} + 3\sin x - e^x\right) dx \qquad\qquad \text{switching to the } power\ form;$$

$$= \int \left(4x^{-2} + 3\sin x - e^x\right) dx \qquad\qquad \text{by the } integration\ rules;$$

$$= 4\int x^{-2}\,dx + 3\int \sin x\,dx - \int e^x\,dx$$

$$\text{using the } basic\ integrals \text{ (see Examples 1.2 and Appendix A);}$$

$$= -\frac{4}{x} - 3\cos x - e^x + C.$$

1.2. Definite Integral

1.2.1. *Definitions*

Definition 1.3 (Partition, Regular Partition).
A set of $n+1$ points $\{x_0, x_1, \ldots, x_n\}$ $(n = 1, 2, \ldots)$ in a *closed* and *bounded* interval $[a, b]$ such that

$$a = x_0 < x_1 < \cdots < x_{n-1} < x_n = b$$

is called a *partition* of $[a, b]$.
The points $x_0, x_1 \ldots, x_n$, called the *grid points*, partition $[a, b]$ into n subintervals

$$[x_0, x_1], [x_1, x_2], \ldots, [x_{n-1}, x_n],$$

the length of the kth *partition interval* being

$$\Delta x_k = x_k - x_{k-1}, \ k = 1, \ldots, n.$$

Remark 1.2. The partition intervals need not be of equal length.

The maximal length of the partition intervals

$$\Delta := \max(\Delta x_1, \Delta x_2, \ldots, \Delta x_n)$$

is called the *diameter* or *norm* of the partition.
If all the partition intervals are of equal length

$$\Delta x = \frac{b-a}{n},$$

the partition $\{x_0, x_1, \ldots, x_n\}$ is called *regular*. In this case, the kth grid point is expressed as follows:

$$x_k = a + k\Delta x, \quad k = 0, 1, \ldots, n.$$

Definition 1.4 (Riemann Sum).
Let a function $f(x)$ be defined on an interval $[a, b]$. Consider a *partition* $\{x_0, x_1, \ldots, x_n\}$ $(n = 1, 2, \ldots)$ of $[a, b]$ $(-\infty < a < b < \infty)$ into n subintervals

$$[x_0, x_1], [x_1, x_2], \ldots, [x_{n-1}, x_n],$$

(of not necessarily equal length), the length of the kth partition interval being

$$\Delta x_k = x_k - x_{k-1}, \ k = 1, \ldots, n.$$

Let x_k^* be an arbitrary *intermediate point* in the kth partition interval $[x_{k-1}, x_k]$, $k = 1, 2, \ldots, n$.
The *Riemann*[1] *sum* for f corresponding to the partition and chosen intermediate points is

$$\sum_{k=1}^{n} f(x_k^*)\Delta x_k = f(x_1^*)\Delta x_1 + f(x_2^*)\Delta x_2 + \cdots + f(x_n^*)\Delta x_n.$$

If the partition is *regular*,

$$\Delta x_k = \Delta x = \frac{b-a}{n}, \ k = 1, \ldots, n,$$

and we have:

$$\sum_{k=1}^{n} f(x_k^*)\Delta x = f(x_1^*)\Delta x + f(x_2^*)\Delta x + \cdots + f(x_n^*)\Delta x.$$

In particular,

---
[1] Bernhard Riemann (1826–1866)

- a *left Riemann sum* $(x_k^* = x_{k-1} = a + (k-1)\Delta x,\ k = 1, 2, \ldots, n)$ is

$$\sum_{k=1}^{n} f(x_{k-1})\Delta x = \sum_{k=1}^{n} f\left(a + (k-1)\Delta x\right)\Delta x;$$

- a *right Riemann sum* $(x_k^* = x_k = a + k\Delta x,\ k = 1, 2, \ldots, n)$ is

$$\sum_{k=1}^{n} f(x_k)\Delta x = \sum_{k=1}^{n} f\left(a + k\Delta x\right)\Delta x;$$

- a *midpoint Riemann sum* $(x_k^* = (x_{k-1} + x_k)/2 = a + (k - 1/2)\Delta x,$ $k = 1, 2, \ldots, n)$ is

$$\sum_{k=1}^{n} f\left((x_{k-1} + x_k)/2\right)\Delta x = \sum_{k=1}^{n} f\left(a + (k - 1/2)\Delta x\right)\Delta x.$$

Definition 1.5 (Net Area).

The *net area* of the region R bounded by the graph of a continuous on an interval $[a, b]$ $(-\infty < a < b < \infty)$ function f and the x-axis is the sum of the areas of the parts of R that lie *above* the x-axis minus the sum of the areas of the parts of R that lie *below* the x-axis.

Definition 1.6 (Definite Integral).

We say that a function f defined on an interval $[a, b]$ $(-\infty < a < b < \infty)$ is *(Riemann-)integrable* on $[a, b]$ if the limit of *Riemann sums*

$$\lim_{\Delta \to 0} \sum_{k=1}^{n} f(x_k^*)\Delta x_k,$$

where

$$\Delta := \max_{1 \le k \le n} \Delta x_k$$

is the *diameter* (or *norm*) of the partition, *exists* (over all partitions of $[a, b]$ and all choices of the intermediate points x_k^*) and is *finite*.
The limit

$$\int_a^b f(x)\, dx := \lim_{\Delta \to 0} \sum_{k=1}^{n} f(x_k^*)\Delta x_k$$

is called the *definite* (or *Riemann*) *integral of f from a to b* (or *over* $[a, b]$).

The integral notation $\int_a^b f(x)\, dx$, uses the *integral symbol* $\int$ and *differential symbol* dx.
The function f is called the *integrand*, x the *integration variable*, and a and b are called the *lower* and *upper limits of integration*, respectively.
The process of evaluation of a definite integral is called *integration*.

1.2.2. Which Functions Are Integrable?

A *necessary*, but *not sufficient*, condition of the Riemann integrability of a function f on an interval $[a, b]$ $(-\infty < a < b < \infty)$ is its *boundedness* of on $[a, b]$ (see, e.g., [4]).

Theorem 1.2 (Integrable Functions).
A function f bounded and having a finite number of discontinuities on an interval $[a, b]$ $(-\infty < a < b < \infty)$ is integrable on $[a, b]$.

Definition 1.7 (Piecewise Continuous Function).
A function f defined on an interval $[a, b]$ $(-\infty < a < b < \infty)$ is called *piecewise continuous* on $[a, b]$ if it is continuous everywhere on $[a, b]$, except at a finite number of *jump discontinuities*.

Remark 1.3. A function f *continuous* on $[a, b]$ is trivially *piecewise continuous*.

Corollary 1.1 (Piecewise Continuous Functions Are Integrable).
A function $f(x)$ piecewise continuous on $[a, b]$ is integrable on $[a, b]$.

For an *integrable* on $[a, b]$ function $f(x)$,

$$\int_a^b f(x)\, dx = Net\ Area.$$

1.2.3. Properties of Definite Integral (Integration Rules)

Theorem 1.3 (Properties of Definite Integral (Integration Rules)).
If functions $f(x)$ and $g(x)$ are integrable on an interval $[a, b]$, then

1. *for any real c,* $\quad \displaystyle\int_a^b cf(x)\, dx = c \int_a^b f(x)\, dx \quad$ (*Constant Factor Rule*)

2. $\displaystyle\int_a^b [f(x) \pm g(x)]\, dx = \int_a^b f(x)\, dx \pm \int_a^b g(x)\, dx \quad$ (*Sum/Difference Rule*)

3. *for $a < c < b$,* $\quad \displaystyle\int_a^b f(x)\, dx = \int_a^c f(x)\, dx + \int_c^b f(x)\, dx \quad$ (*Additivity*)

Remark 1.4. For definite integral, the are NO *Product/Quotient Rules*.

Definition 1.8 (More General Use of Integral Notation).

- For $b = a$, $\quad \displaystyle\int_a^a f(x)\, dx := 0.$

- For $b < a$, $\displaystyle\int_a^b f(x)\,dx := -\int_b^a f(x)\,dx$.

Remark 1.5. These definitions preserve all the above properties of definite integral. The generalized version of *additivity*

$$\int_a^b f(x)\,dx = \int_a^c f(x)\,dx + \int_c^b f(x)\,dx$$

holds for any triple of numbers a, b, and c regardless of the order, provided the function f is *integrable* between any two of the three points.

1.2.4. *Integration by Definition*

Example 1.5 (Integration by Definition).
Evaluate the integral

$$\int_1^4 (x^2 + 3x - 2)\,dx$$

by the definition using *regular partitions* and right *Riemann sums*.

Solution: For the function $f(x) = x^2 + 3x - 2$ on the interval $[1,4]$, considering *regular partitions* of $[1,4]$ into n subintervals with

$$\Delta x = \frac{4-1}{n} = \frac{3}{n}$$

and choosing intermediate points x_k^* to be the *right endpoints*:

$$x_k^* = x_k = 1 + \frac{3}{n}k, \ \ k = 1, 2, \ldots, n,$$

we have:

$$\int_1^4 (x^2 + 3x - 2)\,dx = \lim_{n\to\infty} \sum_{k=1}^n f(x_k)\Delta x = \lim_{n\to\infty} \sum_{k=1}^n f\left(1 + \frac{3}{n}k\right)\frac{3}{n}$$

$$= \lim_{n\to\infty} \sum_{k=1}^n \left[\left(1 + \frac{3}{n}k\right)^2 + 3\left(1 + \frac{3}{n}k\right) - 2\right]\frac{3}{n}$$

transforming and simplifying algebraically;

$$= \lim_{n\to\infty} \sum_{k=1}^n \left[1 + 2\frac{3}{n}k + \frac{9}{n^2}k^2 + 3 + \frac{9}{n}k - 2\right]\frac{3}{n} = \lim_{n\to\infty} \sum_{k=1}^n \left[\frac{9}{n^2}k^2 + \frac{15}{n}k + 2\right]\frac{3}{n}$$

$$= \lim_{n\to\infty} \sum_{k=1}^n \left[\frac{27}{n^3}k^2 + \frac{45}{n^2}k + \frac{6}{n}\right]$$

by the *summation rules*;

$$= \lim_{n\to\infty} \left[\frac{27}{n^3} \sum_{k=1}^{n} k^2 + \frac{45}{n^2} \sum_{k=1}^{n} k + \frac{6}{n} \sum_{k=1}^{n} 1 \right]$$

by the *special summation formulas* (see, e.g., [1, 6]);

$$= \lim_{n\to\infty} \left[\frac{27}{n^3} \frac{n(n+1)(2n+1)}{6} + \frac{45}{n^2} \frac{n(n+1)}{2} + \frac{6}{n} n \right] \quad \text{dividing termwise;}$$

$$= \lim_{n\to\infty} \left[\frac{9}{2} \left(1 + \frac{1}{n}\right) \left(2 + \frac{1}{n}\right) + \frac{45}{2} \left(1 + \frac{1}{n}\right) + 6 \right] \quad \text{by the *limit laws*;}$$

$$= \frac{9}{2}(1+0)(2+0) + \frac{45}{2}(1+0) + 6 = \frac{9}{2}\cdot 2 + \frac{45}{2}\cdot 1 + 6 = 15 + \frac{45}{2} = \frac{75}{2}.$$

1.2.5. Integral Mean Value Theorem

Theorem 1.4 (Integral Mean Value Theorem).
If f is a continuous function on an interval $[a,b]$ ($-\infty < a < b < \infty$), then there exists a point $c \in [a,b]$ such that

$$f(c) = \frac{1}{b-a} \int_a^b f(x)\,dx$$

or

$$f(c)(b-a) = \int_a^b f(x)\,dx.$$

Proof. By the *Extreme Value Theorem* (see, e.g., [1, 6]), the continuous function f attains its *absolute minimum m* and *maximum M* on $[a,b]$. Then for an arbitrary Riemann sum

$$m(b-a) = m\sum_{k=1}^{n} \Delta x_k \leq \sum_{k=1}^{n} f(x_k^*)\Delta x_k \leq M\sum_{k=1}^{n} \Delta x_k = M(b-a).$$

Passing to the limit as $\Delta := \max_{1\leq k\leq n} \Delta x_k \to 0$, we obtain

$$m(b-a) \leq \int_a^b f(x)\,dx \leq M(b-a).$$

Whence,

$$m \leq \frac{1}{b-a} \int_a^b f(x)\,dx \leq M.$$

Since, by the *Intermediate Value Theorem* (see, e.g., [1, 6]), the continuous function f takes on every value between its *absolute minimum* m and *maximum* M on [a,b], there exists a point $c \in [a, b]$ such that

$$f(c) = \frac{1}{b-a} \int_a^b f(x)\,dx.$$

□

Remarks 1.6 (Integral Mean Value Theorem).

- The *Integral Mean Value Theorem* has the following natural *geometric interpretation*.

  If $f(x)$ is a continuous function on an interval $[a, b]$, then there is a point $c \in [a, b]$ such that the net area of the region bounded by the graph of $f(x)$ and the x-axis over $[a, b]$ is equal to the area of the *rectangle* with *base* $[a, b]$ and *height* $f(c)$ provided $f(c) \geq 0$ or the *negative* of the area of the *rectangle* with *base* $[a, b]$ and *height* $-f(c)$ provided $f(c) < 0$.

  If $f(x) \geq 0$ on $[a, b]$, the *net area* is the proper *area*.

- The *Integral Mean Value Theorem* remains valid for $b < a$, in which case, $f(x)$ is required to be continuous *between* a and b where the point c is to be found as well.

1.2.6. *Fundamental Theorem of Calculus*

Definition 1.9 (Area Function).

Let $f(x)$ be a function defined on an interval I and *integrable* on every subinterval $[a, b] \subseteq I$. Fixing a point a in I, we can define the *area function* for $f(x)$ with lower limit a as follows:

$$A(x) := \int_a^x f(t)\,dt, \quad x \in I.$$

For $x \geq a$, $A(x)$ represents the *net area* of the region bounded by the graph of $f(x)$ and the x-axis between a and x and, for $x < a$, considering that

$$A(x) = \int_a^x f(t)\,dt = -\int_x^a f(t)\,dt,$$

$A(x)$ represents the *negative* of the *net area* of the region bounded by the graph of $f(x)$ and the x-axis between a and x.

Theorem 1.5 (Fundamental Theorem of Calculus (Part 1)).
If f is a continuous function on an interval I containing a point a, then the area function for f with lower limit a

$$A(x) := \int_a^x f(t)\,dt, \ x \in I,$$

is differentiable on I and

$$A'(x) = \frac{d}{dx} \int_a^x f(t)\,dt = f(x), \ x \in I.$$

The derivatives at the endpoints (if any) are understood as one-sided.

Proof. For any $x \in I$ and all sufficiently small increments h, such that $x + h \in I$,

$$\frac{A(x+h) - A(x)}{h} = \frac{1}{h} \left[\int_a^{x+h} f(t)\,dt - \int_a^x f(t)\,dt \right]$$

$$\text{by } additivity \text{ (Theorem 1.3);}$$

$$= \frac{1}{h} \int_x^{x+h} f(t)\,dt$$

by the *Integral Mean Value Theorem*, there is a $c(h)$ between x and $x + h$;

$$= f(c(h)).$$

Hence, by the definition of derivative, for any $x \in I$,

$$A'(x) = \lim_{h \to 0} \frac{A(x+h) - A(x)}{h} = \lim_{h \to 0} f(c(h))$$

$$\text{since } c(h) \to x \text{ as } h \to 0, \text{ by the } continuity \text{ of } f(x);$$

$$= f(x).$$

$\square$

Theorem 1.6 (Fundamental Theorem of Calculus (Part 2)).
If a function f is (Riemann-)integrable and has an antiderivative $F(x)$ on an interval $[a, b]$ $(-\infty < a < b < \infty)$, then

$$\int_a^b f(x)\,dx = F(b) - F(a) \qquad (Newton\text{-}Leibniz\ Formula).$$

Proof. Consider an arbitrary partition $\{x_0, x_1, \ldots, x_n\}$ of $[a, b]$ into n subintervals ($n = 1, 2, \ldots$).

Since $F(x)$ satisfies the conditions of the *Mean Value Theorem* on $[a, b]$ (see, e.g., [1, 6]), and thus, on each partition interval $[x_{k-1}, x_k]$, $k = 1, 2, \ldots, n$, for each $k = 1, 2, \ldots, n$, there is a point $x_k^* \in (x_{k-1}, x_k)$ such that

$$F(x_k) - F(x_{k-1}) = F'(x_k^*)(x_k - x_{k-1}) = f(x_k^*)\Delta x_k.$$

Hence,

$$F(b) - F(a) = \sum_{k=1}^{n} [F(x_k) - F(x_{k-1})] = \sum_{k=1}^{n} f(x_k^*)\Delta x_k$$

and

$$F(b) - F(a) = \lim_{\Delta \to 0} \sum_{k=1}^{n} f(x_k^*)\Delta x_k = \int_a^b f(x)\, dx,$$

where $\Delta := \max_{1 \le k \le n} \Delta x_k$.   □

Remarks 1.7 (Fundamental Theorem of Calculus (Part 2)).

- Using the shorthand $F(x)\Big|_a^b$ for $F(b) - F(a)$, we can rewrite the *Newton*[2]-*Leibniz*[3] *Formula* as follows:

$$\int_a^b f(x)\, dx = F(x)\Big|_a^b.$$

- The *Newton-Leibniz Formula* also works when $a \ge b$. Indeed, in this case,

$$\int_a^b f(x)\, dx = -\int_b^a f(x)\, dx = -[F(a) - F(b)] = F(b) - F(a) = F(x)\Big|_a^b.$$

- For a *continuous* function f on an interval $[a, b]$ both conditions of the *Fundamental Theorem of Calculus (Part 2)* are met, and thus, the *Newton-Leibniz Formula* applies.
- Generally, a function f can be integrable without having an antiderivative on $[a, b]$ and need not be integrable while having an antiderivative on $[a, b]$.

---

[2] Isaac Newton (1642–1727)
[3] Gottfried Wilhelm (von) Leibniz (1646–1716)

For instance, the *sign function*

$$\operatorname{sgn} x := \begin{cases} -1 & \text{if } x < 0 \\ 0 & \text{if } x = 0 \\ 1 & \text{if } x > 0 \end{cases}$$

is integrable on $[-1, 1]$, being *piecewise continuous*, but has no anti-derivative on $[-1, 1]$ (see Exercise 1.1).

Whereas, the function

$$f(x) = \begin{cases} 2x \sin \dfrac{1}{x^2} - \dfrac{2}{x} \sin \dfrac{1}{x^2} & \text{if } x \neq 0 \\ 0 & \text{if } x = 0 \end{cases}$$

is not integrable on $[-1, 1]$, being *unbounded*, but has an antiderivative

$$F(x) = \begin{cases} x^2 \sin \dfrac{1}{x^2} & \text{if } x \neq 0 \\ 0 & \text{if } x = 0 \end{cases}$$

(see [4]).

1.2.7. *Total Change Theorem*

An immediate consequence of the *Fundamental Theorem of Calculus* (Part 2) is the following

Theorem 1.7 (Total Change Theorem).
If a function f is a continuously differentiable function on an interval $[a, b]$ $(-\infty < a < b < \infty)$, then

$$\int_a^b f'(x)\, dx = f(b) - f(a) \qquad \text{(Total Change Formula)},$$

i.e., the integral of the instantaneous rate of change function f' over the interval $[a, b]$ is equal to the total change of f over $[a, b]$.

Remark 1.8. The *Total Change Theorem* has the following *physical interpretation*.

If v is the *(continuous) velocity function* of an object moving along the x-axis with the *position function* x, then the *displacement* of the object between the moments $t = a$ and $t = b$ is

$$x(b) - x(a) = \int_a^b v(t)\, dt,$$

which follows immediately from the prior theorem in view of $v(t) = x'(t)$, $a \leq t \leq b$.

1.2.8. *Integrals of Even and Odd Functions*

Theorem 1.8 (Integrals of Symmetric Functions over Symmetric Intervals).
Let $a > 0$ and f be an integrable function on the interval $[-a, a]$.

- *If f is even $(f(-x) = f(x))$,* $\displaystyle\int_{-a}^{a} f(x)\, dx = 2 \int_{0}^{a} f(x)\, dx.$
- *If f is odd $(f(-x) = -f(x))$,* $\displaystyle\int_{-a}^{a} f(x)\, dx = 0.$

Example 1.6 (Integrals of Symmetric Functions over Symmetric Intervals).
Evaluate the integral

$$\int_{-4}^{4} \frac{\sin^7 x}{(x^2 + 1)^{10}}\, dx.$$

Solution: Since, the *continuous function* $\dfrac{\sin^7 x}{(x^2 + 1)^{10}}$ is *odd* and the interval $[-4, 4]$ is *symmetric* about 0,

$$\int_{-4}^{4} \frac{\sin^7 x}{(x^2 + 1)^{10}}\, dx = 0.$$

Chapter 2

Direct Integration

Recall that the process of finding an indefinite integral is called *integration*. Just like for differentiation, for integration purposes, along with the *integration rules* (see Theorem 1.1), we need a set of validated or established basic integrals, called the *table of basic integrals*.

2.1. Table Integrals and Useful Integration Formula

For our integration purposes, we use the *Table of Basic Integrals* consisting of twenty-one integrals (see Appendix A). We have already validated most of them in the preceding examples. Observe that there exist much more extensive tables of integrals.

Exercise 2.1 (Validating Table Integrals).
Validate integrals 12 and 14 of the *Table of Basic Integrals* (Appendix A).

We are almost ready to integrate now. However, at the moment, we are in the situation of an awkward deficiency demonstrated by the following examples: knowing the table integrals

$$\int e^x \, dx = e^x + C \quad \text{or} \quad \int \sin x \, dx = -\cos x + C,$$

we do not know how to evaluate their slightest variations such as

$$\int e^{-x} \, dx \quad \text{or} \quad \int \sin(2x + 3) \, dx.$$

These are particular cases of the routine situation of evaluating the integral of the form

$$\int f(ax + b) \, dx,$$

where a and b are real coefficients with $a \neq 0$, when $\int f(x)\,dx$ is known, which, as we shall see in Sec. 3.1.4, can be treated by the *Method of Substitution*.

To deal with this very common situation, let us, however, not wait until the *Substitution Method* is developed, but prove and start applying immediately the following

Theorem 2.1 (Useful Integration Formula).
Let a and b be real coefficients with $a \neq 0$. If

$$\int f(x)\,dx = F(x) + C,$$

then

$$\int f(ax+b)\,dx = \frac{1}{a}F(ax+b) + C.$$

Proof. Considering that $F'(x) = f(x)$, by the *Chain Rule* (see, e.g., [1, 6]), we have:

$$\frac{d}{dx}\left[\frac{1}{a}F(ax+b)\right] = \frac{1}{a}F'(ax+b)[ax+b]' = \frac{1}{a}f(ax+b)a = f(ax+b).$$

$\square$

Remarks 2.1 (Useful Integration Formula).

- We intentionally do not indicate the intervals of integration not to obscure the simplicity of the formulation and proof.
- As we see in Sec. 3.1.4, the *Useful Integration Formula* (Theorem 2.1) is a special case of the so-called *trivial substitution*.

Examples 2.1 (Applying the Useful Integration Formula). Hereafter, the boxes contain explanatory formulas and/or text.

1. $\int e^{-x}\,dx$ by the *useful integration formula* with $a = -1$ and $b = 0$;

$$= \frac{1}{(-1)}e^{-x} + C = -e^{-x} + C.$$

2. $\int \sin(2x+3)\,dx$

by the *useful integration formula* with $a = 2$ and $b = 3$;

$$= \frac{1}{2}(-\cos(2x+3)) + C = -\frac{1}{2}\cos(2x+3) + C.$$

3. $\displaystyle\int \sqrt[3]{4-10x}\,dx$ switching to the *power form*;

$$= \int (4-10x)^{1/3}\,dx$$

by the *useful integration formula* with $a = -10$ and $b = 4$;

$$= \frac{1}{(-10)} \cdot \frac{1}{4/3}(4-10x)^{4/3} + C = -\frac{3}{40}(4-10x)^{4/3} + C.$$

4. $\displaystyle\int \frac{1}{x/2 + \pi}\,dx$

by the *useful integration formula* with $a = 1/2$ and $b = \pi$;

$$= \frac{1}{1/2}\ln|x/2 + \pi| + C = 2\ln|x/2 + \pi| + C.$$

Table of Basic Integrals (Appendix A), integrals 17 and 21.

For $a > 0$,

5. $\displaystyle\int \frac{1}{x^2 + a^2}\,dx = \frac{1}{a^2}\int \frac{1}{(x/a)^2 + 1}\,dx$

by the *useful integration formula* with $a = 1/a$ and $b = 0$;

$$= \frac{1}{a^2}\frac{1}{1/a}\arctan\frac{x}{a} + C = \frac{1}{a}\arctan\frac{x}{a} + C \quad \text{on } (-\infty, \infty).$$

6. $\displaystyle\int \frac{1}{x\sqrt{x^2 - a^2}}\,dx = \frac{1}{a^2}\int \frac{1}{(x/a)\sqrt{(x/a)^2 - 1}}\,dx + C$

by the *useful integration formula* with $a = 1/a$ and $b = 0$;

$$= \frac{1}{a^2}\frac{1}{1/a}\operatorname{arcsec}\left|\frac{x}{a}\right| + C = \frac{1}{a}\operatorname{arcsec}\left|\frac{x}{a}\right| + C$$

on each of the intervals $(-\infty, -a)$, (a, ∞).

Exercise 2.2 (Deriving Table Integrals).

Apply the *Useful Integration Formula* (Theorem 2.1) to derive integral 19 of the *Table of Basic Integrals* (Appendix A).

When evaluating an integral of the form

$$\int (ax + b)^n \, dx,$$

where a and b are real coefficients and $n = 1, 2, 3, \ldots$ is a natural number, the *Useful Integration Formula* (Theorem 2.1) becomes especially handy for large values of the exponent n, when the use of the *binomial formula* leads to cumbersome computations.

Example 2.2 (Applying the Useful Integration Formula).
Evaluate the integral

$$\int (5x - 7)^{99} \, dx.$$

Solution:　Using the *binomial formula* (see Appendix C) to expand $(5x - 7)^{99}$ would result into 100 terms. Applying the *Useful Integration Formula* (Theorem 2.1) "saves the day":

$$\int (5x - 7)^{99} \, dx$$

by the *useful integration formula* with $a = 5$ and $b = -7$;

$$= \frac{1}{5} \frac{1}{100} (5x - 7)^{100} + C = \frac{1}{500} (5x - 7)^{100} + C.$$

2.2.　What Is Direct Integration and How Does It Work?

What Is Direct Integration?

Definition 2.1 (Direct Integration).
By *direct integration*, we understand the process of integration, which, using the *integration rules* (Theorem 1.1) alone, reduces the integral of a given function to a combination of table integrals. Such integration makes no use of any special integration techniques, but may employ the *Useful Integration Formula* (Theorem 2.1) when appropriate.

Remark 2.2. When executing direct integration, before applying the *integration rules*, one may need to manipulate the integrand performing multiplication/division, applying relevant identities of algebra/trigonometry, or using such tricks as transforming products into sums, multiplying and dividing by the conjugate radical expression, or completing the square.

How Does Direct Integration Work?

In the following examples, we consider various scenarios of *direct integration* explaining its execution step-by-step.

2.2.1. *By Integration Rules Only*

Example 2.3 (Integration by the Rules Only).

$$\int (7x^3 + 5e^{-x} + 14)\,dx \qquad\qquad \text{by the } integration\ rules;$$

$$= 7\int x^3\,dx + 5\int e^{-x}\,dx + 14\int 1\,dx$$

$$\qquad\qquad \text{by the } useful\ integration\ formula;$$

$$= \frac{7}{4}x^4 - 5e^{-x} + 14x + C.$$

2.2.2. *Multiplication/Division Before Integration*

When evaluating an integral one is to be mindful of the absence of the *product* and *quotient rules* for integration.

Thus, the following "solution"

$$\int \frac{(x-1)(x+2)}{\sqrt{x}}\,dx = \frac{\int (x-1)\,dx \int (x+2)\,dx}{\int \sqrt{x}\,dx} = \cdots$$

is *incorrect*.

To solve correctly, we are to execute multiplication and division before integration.

Examples 2.4 (Multiplication/Division Before Integration).

1. $\displaystyle \int \frac{(x-1)(x+2)}{\sqrt{x}}\,dx$ multiplying and switching to the *power form*;

$$= \int \frac{x^2 + x - 2}{x^{1/2}}\,dx \qquad\qquad \text{dividing termwise;}$$

$$= \int (x^{3/2} + x^{1/2} - 2x^{-1/2})\,dx \qquad\qquad \text{by the } integration\ rules;$$

$$= \int x^{3/2}\,dx + \int x^{1/2}\,dx - 2\int x^{-1/2}\,dx = \frac{2}{5}x^{5/2} + \frac{2}{3}x^{3/2} - 4x^{1/2} + C.$$

2. $\displaystyle\int \frac{x^3 - 2x^2 + 4x - 5}{x - 1}\, dx$

$\qquad\qquad$ dividing $\qquad$ $\boxed{\dfrac{x^3 - 2x^2 + 4x - 5}{x - 1} = x^2 - x + 3 - \dfrac{2}{x - 1}}$ $\qquad$ (*verify*);

$\displaystyle = \int \left(x^2 - x + 3 - \frac{2}{x - 1} \right) dx$ $\qquad\qquad\qquad$ by the *integration rules*;

$\displaystyle = \int x^2 \, dx - \int x \, dx + 3 \int 1 \, dx - 2 \int \frac{1}{x - 1}\, dx$

$\qquad\qquad\qquad\qquad\qquad\qquad\qquad$ by the *useful integration formula*;

$\displaystyle = \frac{x^3}{3} - \frac{x^2}{2} + 3x - 2 \ln|x - 1| + C.$

The latter example represents a particular case of integration of rational functions considered in detail in Chapter 7.

2.2.3. *Applying Minor Adjustments*

By *"minor adjustments"* to the integrand, we understand adding and subtracting or multiplying and dividing by the same constant.

Example 2.5 (Applying Minor Adjustments).

$\displaystyle \int x\sqrt{5 - 2x}\, dx$ $\qquad\qquad\qquad\qquad$ multiplying and dividing by -2;

$\displaystyle = -\frac{1}{2} \int (-2x)(5 - 2x)^{1/2}\, dx$ $\qquad\qquad\qquad$ adding and subtracting 5;

$\displaystyle = -\frac{1}{2} \int (5 - 2x - 5)(5 - 2x)^{1/2}\, dx = -\frac{1}{2} \int \left[(5 - 2x)^{3/2} - 5(5 - 2x)^{1/2} \right] dx$

$\qquad\qquad\qquad\qquad\qquad\qquad\qquad$ by the *integration rules*;

$\displaystyle = -\frac{1}{2} \int (5 - 2x)^{3/2}\, dx + \frac{5}{2} \int (5 - 2x)^{1/2}\, dx$

$\qquad\qquad\qquad\qquad\qquad\qquad\qquad$ by the *useful integration formula*;

$\displaystyle = -\frac{1}{2} \frac{1}{(-2)} \frac{2}{5} (5 - 2x)^{5/2} + \frac{5}{2} \frac{1}{(-2)} \frac{2}{3} (5 - 2x)^{3/2} + C$

$\displaystyle = \frac{1}{10} (5 - 2x)^{5/2} - \frac{5}{6} (5 - 2x)^{3/2} + C.$

2.2.4.   *Using Identities*

Certain identities of algebra and trigonometry useful for integration can be found in Appendix C.

Examples 2.6 (Using Identities).

1. $\displaystyle\int 2^{5x}3^{-x}\,dx$ by the *exponents laws* (see Appendix C);

$$= \int \left(2^5 \cdot 3^{-1}\right)^x dx = \int \left(\frac{32}{3}\right)^x dx = \frac{(32/3)^x}{\ln(32/3)} + C.$$

2. $\displaystyle\int \frac{\cos x + 1}{\sin^2 x}\,dx$ dividing termwise and rewriting equivalently;

$$= \int \left[\frac{\cos x}{\sin^2 x} + \frac{1}{\sin^2 x}\right] dx = \int \left[\frac{1}{\sin x}\frac{\cos x}{\sin x} + \left(\frac{1}{\sin x}\right)^2\right] dx$$

$$\text{since} \quad \boxed{\frac{1}{\sin x} = \csc x, \quad \frac{\cos x}{\sin x} = \cot x} \quad \text{(see Appendix C)};$$

$$= \int \left[\csc x \cot x + \csc^2 x\right] dx \qquad\qquad \text{by the *integration rules*};$$

$$= \int \csc x \cot x\,dx + \int \csc^2 x\,dx = -\csc x - \cot x + C.$$

When evaluating the trigonometric integral

$$\int \cos^2 x\,dx,$$

in the absence of the *product rule* for integration, we are to use *power-reduction identity*

$$\cos^2 \theta = \frac{1 + \cos 2\theta}{2}$$

(see Appendix C) as follows:

Example 2.7 (Using Identities).

$$\int \cos^2 x\,dx$$

       by the *power-reduction identity* with $\theta = x$, $\boxed{\cos^2 x = \dfrac{1 + \cos 2x}{2}}$;

$$= \int \frac{1 + \cos 2x}{2} \, dx \qquad \qquad \text{by the } \textit{integration rules;}$$

$$= \frac{1}{2} \left[\int 1 \, dx + \int \cos 2x \, dx \right] \qquad \text{by the } \textit{useful integration formula;}$$

$$= \frac{1}{2} \left[x + \frac{1}{2} \sin 2x \right] + C = \frac{1}{2} x + \frac{1}{4} \sin 2x + C.$$

More generally, using a relevant *power-reduction identity* allows us to evaluate integrals of the form

$$\int \cos^2 (ax + b) \, dx, \quad \int \sin^2 (ax + b) \, dx,$$

where a and b are real coefficients with $a \neq 0$.

Similarly, the trigonometric integral

$$\int \sin 3x \cos x \, dx$$

containing a product is found via the *product-to-sum identity*

$$\sin \alpha \cos \beta = \frac{1}{2} \left[\sin(\alpha - \beta) + \sin(\alpha + \beta) \right]$$

(see Appendix C) as follows:

Example 2.8 (Using Identities).

$$\int \sin 3x \cos x \, dx$$

by the relevant *product-to-sum identity* with $\alpha = 3x$ and $\beta = x$,

$$\boxed{\sin 3x \cos x = \frac{1}{2} \left[\sin(3x - x) + \sin(3x + x) \right] = \frac{1}{2} \left[\sin 2x + \sin 4x \right]};$$

$$= \int \frac{1}{2} \left[\sin 2x + \sin 4x \right] \, dx \qquad \text{by the } \textit{integration rules;}$$

$$= \frac{1}{2} \left[\int \sin 2x \, dx + \int \sin 4x \, dx \right] \qquad \text{by the } \textit{useful integration formula;}$$

$$= \frac{1}{2} \left[\frac{1}{2} (-\cos 2x) + \frac{1}{4} (-\cos 4x) \right] + C = -\frac{1}{4} \cos 2x - \frac{1}{8} \cos 4x + C.$$

More generally, this approach applies to the integrals of the form

$$\int \cos(ax + b)\cos(cx + d)\,dx,$$

$$\int \sin(ax + b)\sin(cx + d)\,dx,$$

$$\int \sin(ax + b)\cos(cx + d)\,dx,$$

where a, b, c, and d are real coefficients with $a, c \neq 0$.

The following trigonometric integral

$$\int \tan^2 5x\,dx$$

can be evaluated via the *Pythagorean identity*

$$\tan^2 \theta = \sec^2 \theta - 1$$

(see Appendix C) as follows:

Example 2.9 (Using Identities).

$$\int \tan^2 5x\,dx$$

$$\text{by the latter identity with } \theta = 5x, \quad \boxed{\tan^2 5x = \sec^2 5x - 1};$$

$$= \int \left[\sec^2 5x - 1\right]\,dx \qquad\qquad\qquad \text{by the *integration rules*;}$$

$$= \int \sec^2 5x\,dx - \int 1\,dx \qquad\qquad \text{by the *useful integration formula*;}$$

$$= \frac{1}{5}\tan 5x - x + C.$$

More generally, this approach applies to the integrals of the form

$$\int \tan^2(ax + b)\,dx, \quad \int \cot^2(ax + b)\,dx,$$

where a and b are real coefficients with $a \neq 0$.

2.2.5.   *Transforming Products into Sums*

As is seen from Example 2.8, in the absence of the product rule for integration, one can attempt transforming a product into a sum. Let us use this approach for the following problem.

Example 2.10 (Transforming Products into Sums).
Evaluate the integral

$$\int \frac{1}{(2x-1)(2x+3)}\,dx.$$

Solution: The integrand is a product

$$\frac{1}{2x-1}\cdot\frac{1}{2x+3}.$$

However, the simple fact that the polynomials $2x-1$ and $2x+3$ share the same nonconstant term $2x$ allows us to rewrite it as follows:

$$\frac{1}{2x-1}\cdot\frac{1}{2x+3}=\frac{1}{4}\left[\frac{1}{2x-1}-\frac{1}{2x+3}\right]$$

by replacing the product with the difference

$$\frac{1}{2x-1}-\frac{1}{2x+3}$$

and scaling the latter by coefficient $1/4$, where the denominator 4 is the difference of the constant terms: $3-(-1)$.
Thus,

$$\int \frac{1}{(2x-1)(2x+3)}\,dx \qquad\qquad \text{transforming product into sum:}$$

$$\boxed{\frac{1}{(2x-1)(2x+3)}=\frac{1}{4}\left[\frac{1}{2x-1}-\frac{1}{2x+3}\right]};$$

$$=\int \frac{1}{4}\left[\frac{1}{2x-1}-\frac{1}{2x+3}\right]dx \qquad\qquad \text{by the \textit{integration rules};}$$

$$=\frac{1}{4}\left[\int \frac{1}{2x-1}\,dx-\int \frac{1}{2x+3}\,dx\right] \qquad \text{by the \textit{useful integration formula};}$$

$$=\frac{1}{4}\left[\frac{1}{2}\ln|2x-1|-\frac{1}{2}\ln|2x+3|\right]+C=\frac{1}{8}\left[\ln|2x-1|-\ln|2x+3|\right]+C.$$

The prior problem, as well as problem 2 from Examples 2.4, is a special case of integration of rational functions studied in Chapter 7.

Example 2.11 (Table of Basic Integrals (Appendix A), integral 18).
For $a>0$,

$$\int \frac{1}{x^2-a^2}\,dx=\frac{1}{2a}\ln\left|\frac{x-a}{x+a}\right|+C$$

on each of the intervals $(-\infty, -a)$, $(-a, a)$, (a, ∞).

$$\int \frac{1}{x^2 - a^2}\, dx \qquad\qquad \text{transforming product into sum:}$$

$$\boxed{\frac{1}{x^2 - a^2} = \frac{1}{(x-a)(x+a)} = \frac{1}{2a}\left[\frac{1}{x-a} - \frac{1}{x+a}\right]};$$

$$= \int \frac{1}{2a}\left[\frac{1}{x-a} - \frac{1}{x+a}\right] dx \qquad\qquad \text{by the } \textit{integration rules};$$

$$= \frac{1}{2a}\left[\int \frac{1}{x-a}\, dx - \int \frac{1}{x+a}\, dx\right]$$

$$\text{by the } \textit{useful integration formula} \text{ with } a = 1 \text{ and } b = \pm a;$$

$$= \frac{1}{2a}\left[\ln|x-a| - \ln|x+a|\right] + C$$

$$\text{by the } \textit{laws of logarithms} \text{ (see Appendix C)};$$

$$= \frac{1}{2a}\ln\left|\frac{x-a}{x+a}\right| + C.$$

2.2.6. *Using Conjugate Radical Expressions*

Example 2.12 (Using Conjugate Radical Expressions).

$$\int \frac{1}{\sqrt{2x+7} - \sqrt{2x+3}}\, dx$$

$$\text{multiplying and dividing by the } \textit{conjugate radical expression};$$

$$= \int \frac{\sqrt{2x+7} + \sqrt{2x+3}}{(\sqrt{2x+7} - \sqrt{2x+3})(\sqrt{2x+7} + \sqrt{2x+3})}\, dx \qquad \text{simplifying};$$

$$= \int \frac{\sqrt{2x+7} + \sqrt{2x+3}}{2x+7 - (2x+3)}\, dx = \int \frac{1}{4}\left[\sqrt{2x+7} + \sqrt{2x+3}\right] dx$$

$$\text{switching to the } \textit{power form};$$

$$= \int \frac{1}{4}\left[(2x+7)^{1/2} + (2x+3)^{1/2}\right] dx \qquad \text{by the } \textit{integration rules};$$

$$= \frac{1}{4}\left[\int (2x+7)^{1/2}\, dx + \int (2x+3)^{1/2}\, dx\right]$$

$$\text{by the } \textit{useful integration formula};$$

$$= \frac{1}{4}\frac{1}{2}\frac{2}{3}(2x+7)^{3/2} + \frac{1}{4}\frac{1}{2}\frac{2}{3}(2x+3)^{3/2} + C = \frac{(2x+7)^{3/2}}{12} + \frac{(2x+3)^{3/2}}{12} + C.$$

2.2.7. *Square Completion*

The *square completion* technique is applied when evaluating integrals containing *quadratic polynomials* of the following two types:

$$\int \frac{1}{ax^2 + bx + c}\, dx \quad \text{or} \quad \int \frac{1}{\sqrt{ax^2 + bx + c}}\, dx,$$

where a, b, and c are numeric coefficients with $a \neq 0$.

Example 2.13 (Square Completion).

$$\int \frac{1}{3x^2 - x + 4}\, dx \qquad\qquad \text{factoring out the *leading coefficient*;}$$

$$= \frac{1}{3} \int \frac{1}{x^2 - x/3 + 4/3}\, dx$$

completing the square:

$$\boxed{x^2 - \frac{1}{3}x + \frac{4}{3} = x^2 - 2\frac{1}{6}x + \frac{1}{36} - \frac{1}{36} + \frac{4}{3} = \left(x - \frac{1}{6}\right)^2 + \frac{47}{36}};$$

$$= \frac{1}{3} \int \frac{1}{(x - 1/6)^2 + (\sqrt{47}/6)^2}\, dx \qquad \text{by the *useful integration formula*;}$$

$$= \frac{1}{3}\frac{1}{\sqrt{47}/6} \arctan \frac{x - 1/6}{\sqrt{47}/6} + C = \frac{2}{\sqrt{47}} \arctan \frac{6x - 1}{\sqrt{47}} + C.$$

Remark 2.3. When evaluating an integral of the form

$$\int \frac{1}{ax^2 + bx + c}\, dx,$$

after completing the square, we always arrive at one of the two possibilities (up to a constant factor):

- $\int \frac{1}{(x + h)^2 + d^2}\, dx = \frac{1}{d} \arctan \frac{x + h}{d} + C$ (as above) or
- $\int \frac{1}{(x + h)^2 - d^2}\, dx = \frac{1}{2d} \ln \left| \frac{(x + h) - d}{(x + h) + d} \right| + C.$

Example 2.14 (Square Completion).

$$\int \frac{1}{\sqrt{1 + 5x - x^2}}\, dx \qquad\qquad \text{*completing the square:*}$$

$$1 + 5x - x^2 = -\left[x^2 - 2\frac{5}{2}x + \frac{25}{4} - \frac{25}{4} - 1\right] = \frac{29}{4} - \left(x - \frac{5}{2}\right)^2;$$

$$= \int \frac{1}{\sqrt{\left(\sqrt{29}/2\right)^2 - (x - 5/2)^2}} \, dx \qquad \text{by the } \textit{useful integration formula};$$

$$= \arcsin \frac{x - 5/2}{\sqrt{29}/2} + C = \arcsin \frac{2x - 5}{\sqrt{29}} + C.$$

Remark 2.4. When evaluating an integral of the form

$$\int \frac{1}{\sqrt{ax^2 + bx + c}} \, dx,$$

after completing the square, we always arrive at one of the two possibilities (up to a constant factor):

- $\int \dfrac{1}{\sqrt{d^2 - (x+h)^2}} \, dx = \arcsin \dfrac{x+h}{d} + C$ (as above) or

- $\int \dfrac{1}{\sqrt{(x+h)^2 \pm d^2}} \, dx = \ln|x + h + \sqrt{(x+h)^2 \pm d^2}| + C.$

2.3. Direct Integration for Definite Integral

Definition 2.2 (Direct Integration).
By *direct integration* for definite integral, we understand the process of integration, which is completely based on the *integration rules* (Theorem 1.3), possibly, the *Useful Integration Formula* (Theorem 2.1), and the *Newton-Leibniz Formula* (Theorem 1.6) and makes no use of special integration methods.

Examples 2.15 (Direct Integration for Definite Integral).

1. $\displaystyle \int_1^8 \sqrt[3]{x} \, dx$ \qquad switching to the *power form*;

$$= \int_1^8 x^{1/3} \, dx \qquad \text{by } \textit{Newton-Leibniz Formula};$$

$$= \frac{3}{4}x^{4/3} \Big|_1^8 = \frac{3}{4}(8^{4/3} - 1^{4/3}) = \frac{3}{4}(16 - 1) = \frac{45}{4}.$$

2. $\displaystyle\int_0^\pi \left(\sin x - 3e^{-x}\right)\,dx$ by the *integration rules*;

$\displaystyle = \int_0^\pi \sin x\,dx - 3\int_0^\pi e^{-x}\,dx$

     by the *useful integration formula* and the *Newton-Leibniz Formula*;

$\displaystyle = -\cos x\Big|_0^\pi + 3e^{-x}\Big|_0^\pi = -(\cos\pi - \cos 0) + 3\left(e^{-\pi} - e^0\right)$

$\displaystyle = -(-1-1) + 3(e^{-\pi} - 1) = 3e^{-\pi} - 1.$

3. $\displaystyle\int_1^2 \frac{(x+1)^2}{x}\,dx$ squaring and dividing termwise;

$\displaystyle = \int_1^2 \frac{x^2 + 2x + 1}{x}\,dx = \int_1^2 \left[x + 2 + \frac{1}{x}\right]dx$

     by the *integration rules* and the *Newton-Leibniz Formula*;

$\displaystyle = \left[\frac{x^2}{2} + 2x + \ln|x|\right]\Big|_1^2 = \frac{4}{2} + 2\cdot 2 + \ln 2 - \frac{1}{2} - 2 - \ln 1 = \frac{7}{2} + \ln 2.$

4. For $f(x) = \begin{cases} -1, & 0 \le x < 1, \\ 2^x, & 1 \le x \le 3 \end{cases}$,   by *additivity* (Theorem 1.3),

$\displaystyle\int_0^3 f(x)\,dx = \int_0^1 (-1)\,dx + \int_1^3 2^x\,dx$

     by the *integration rules* and the *Newton-Leibniz Formula*;

$\displaystyle = -x\Big|_0^1 + \frac{2^x}{\ln 2}\Big|_1^3 = -(1-0) + (2^3 - 2^1)/\ln 2 = -1 + 6/\ln 2.$

5. $\displaystyle\int_{-1}^2 |2t - 1|\,dt$

since $|2t-1| = \begin{cases} -(2t-1), & -1 \le t \le 1/2, \\ 2t-1, & 1/2 \le t \le 2 \end{cases}$, by *additivity* (Theorem 1.3);

$\displaystyle = \int_{-1}^{1/2} [-(2t-1)]\,dt + \int_{1/2}^2 (2t-1)\,dt$

     by the *integration rules* and the *Newton-Leibniz Formula*;

$$= -\int_{-1}^{1/2} (2t - 1) \, dt + \int_{1/2}^{2} (2t - 1) \, dt = -(t^2 - t)\Big|_{-1}^{1/2} + (t^2 - t)\Big|_{1/2}^{2}$$

$$= -[(1/4 - 1/2) - (1 - (-1))] + [(4 - 2) - (1/4 - 1/2)]$$

$$= 1/4 + 2 + 2 + 1/4 = 9/2.$$

2.4. Applications

Examples 2.16 (Applications).

1. Find the solution of the *initial value problem*

$$y' = x - \frac{1}{x}, \ y(-1) = 1.$$

Solution: Integrating by the rules only, we find:

$$y = \int \left(x - \frac{1}{x}\right) dx = \int x \, dx - \int \frac{1}{x} \, dx = \frac{x^2}{2} - \ln|x| + C$$

on each of the intervals $(-\infty, 0)$, $(0, \infty)$, where C is an arbitrary real constant. Since the *initial condition* is given at $x = -1$, the interval of interest for us is $(-\infty, 0)$.

Hence, all possible solutions (the *general solution*) of the differential equation on $(-\infty, 0)$ are given by the formula

$$y = \frac{x^2}{2} - \ln|x| + C$$

with an arbitrary real C.

We find the solution of the initial value problem by plugging in $x = -1$ and solving the obtained equation for C:

$$y(-1) = \frac{(-1)^2}{2} - \ln|-1| + C = 1 \Leftrightarrow 1/2 + C = 1 \Leftrightarrow C = 1/2.$$

Hence, the desired solution is

$$y = x^2/2 - \ln|x| + 1/2, \ x < 0.$$

2. Find the *volume* of the solid of generated by revolving the region bounded by

$$y = \sqrt{x} \quad \text{and} \quad y = x^3$$

(a) about the *x-axis*,
(b) about the *y-axis*.

Solution: Let us first find where the graphs intersect by solving the equation

$$\sqrt{x} = x^3 \Leftrightarrow x^3 - x^{1/2} = 0 \Leftrightarrow x^{1/2}(x^{5/2} - 1) = 0 \Leftrightarrow x = 0 \text{ or } x = 1.$$

Remark 2.5. Had we divided through by $\sqrt{x}$, we would have lost the solution $x = 0$.

Hence, the region is bounded by the graphs of $y = \sqrt{x}$ and $y = x^3$ over the interval $[0, 1]$. Observe that

$$\sqrt{x} \geq x^3 \text{ on } [0, 1].$$

(a) Since the rotation axis is the axis of definition, we apply the *washer method* (see, e.g., [1, 6]):

$$V = \int_0^1 \pi \left[(\sqrt{x})^2 - (x^3)^2 \right] dx = \int_0^1 \pi \left[x - x^6 \right] dx$$

by the *integration rules* and the *Newton-Leibniz Formula*;

$$= \pi \int_0^1 \left[x - x^6 \right] dx = \pi \left[\frac{x^2}{2} - \frac{x^7}{7} \right]\Big|_0^1 = \pi \left[\frac{1^2}{2} - \frac{1^7}{7} \right] = \frac{5\pi}{14} \text{ un.}^3.$$

(b) Since the rotation axis is perpendicular to the axis of definition, we apply the *shell method* (see, e.g., [1, 6]):

$$V = \int_0^1 2\pi x \left[\sqrt{x} - x^3 \right] dx$$

switching to the *power form* and multiplying;

$$= \int_0^1 2\pi x \left[x^{1/2} - x^3 \right] dx = \int_0^1 2\pi \left[x^{3/2} - x^4 \right] dx$$

by the *integration rules* and the *Newton-Leibniz Formula*;

$$= 2\pi \int_0^1 \left[x^{3/2} - x^4 \right] dx = 2\pi \left[\frac{2}{5} x^{5/2} - \frac{x^5}{5} \right]\Big|_0^1 = 2\pi \left[\frac{2}{5} 1^{5/2} - \frac{1^5}{5} \right]$$

$$= \frac{2\pi}{5} \text{ un.}^3.$$

2.5. Practice Problems

Evaluate the integrals.

1. $\int \dfrac{1}{\sqrt{5x-6}}\,dx$

2. $\int (2-3x)^{10}\,dx$

3. $\int \left(3x^2+2x-\dfrac{1}{x}\right)dx$

4. $\int e^x\left(1-\dfrac{e^{-x}}{x^2}\right)dx$

5. $\int \dfrac{x^2+2}{x^4}\,dx$

6. $\int \dfrac{2x^2+x+1}{x-1}\,dx$

7. $\int \dfrac{(\sqrt{x}-1)^3}{x}\,dx$

8. $\int x\sqrt[3]{2x-1}\,dx$

9. $\int \dfrac{3x+2}{2x+3}\,dx$

10. $\int \dfrac{x^4}{x^2+2}\,dx$

11. $\int \dfrac{3x^3-5x+2}{x+2}\,dx$

12. $\int (2^x+3^x)^2\,dx$

13. $\int \dfrac{\sqrt{x^4+x^{-4}+2}}{x^3}\,dx$

14. $\int \dfrac{\sqrt{1+x^2}+\sqrt{1-x^2}}{\sqrt{1-x^4}}\,dx$

15. $\int \dfrac{1}{\sqrt{x+4}+\sqrt{x}}\,dx$

16. $\int \dfrac{1}{x^2+2x-3}\,dx$

17. $\int \dfrac{1}{4x^2+4x+3}\,dx$

18. $\int \dfrac{1}{\sqrt{5-4x-x^2}}\,dx$

19. $\int \dfrac{1}{\sqrt{3x^2-2x-1}}\,dx$

20. $\int \dfrac{1-\cos^3 x}{\cos^2 x}\,dx$

21. $\int \dfrac{\cos 2x}{\sin^2 x\cos^2 x}\,dx$

22. $\int (\cos x+\sin x)^2\,dx$

23. $\int \sin^2(x/2+1)\,dx$

24. $\int \cos 5x\cos 3x\,dx$

25. $\int \dfrac{1}{1+\cos x}\,dx$

26. $\int \sqrt{1+\sin 2x}\,dx$

27. $\int_1^2 (x-1)^6\,dx$

28. $\int_1^4 \dfrac{1-\sqrt{x}}{x}\,dx$

29. $\int_0^1 \dfrac{x^3+1}{x+1}\,dx$

30. $\int_{\pi/4}^{\pi/2} \cot^2 x\,dx$

Chapter 3

Method of Substitution

3.1. Substitution for Indefinite Integral

3.1.1. *What for? Why? How?*

What Is the Method for?

The *Method of Substitution* for indefinite integral is used for finding integrals of the form:

$$\int f(g(x))g'(x)\,dx,$$

e.g.,

$$\int 10^{\sin x} \cos x\,dx$$

with $f(u) = 10^u$ and $g(x) = \sin x$.

Why Does the Method Work?

The method is based on the following

Theorem 3.1 (Substitution Rule for Indefinite Integral).
Let

- *a function f have an antiderivative F on an interval I and*
- *a function g be differentiable on an interval J and take values in I.*

Then on the interval J, with the substitution $u = g(x)$,

$$\int f(g(x))g'(x)\,dx = \int f(u)\,du.$$

Proof. Since

$$\int f(u)\,du = F(u) + C \text{ on } I,$$

with the *substitution* $u = g(x)$, we are to show that

$$\int f(g(x))g'(x)\, dx = F(g(x)) + C \text{ on } J,$$

which is readily verified by the *Chain Rule*:

$$\frac{d}{dx}F(g(x)) = F'(g(x))g'(x) = f(g(x))g'(x) \text{ on } J.$$

$\square$

Remark 3.1. Thus, the *Substitution Rule for Indefinite Integral* is the integral form of the *Chain Rule* for differentiation (see, e.g., [1, 6]).

How Does the Method Work?

An integral of the form

$$\int f(g(x))g'(x)\, dx,$$

with functions f and g satisfying the conditions of the *Substitution Rule*, is found in the following *three steps*:

$$\int f(g(x))g'(x)\, dx$$

Step 1. *Substitute* $\boxed{u = g(x),\ \ du = g'(x)\, dx}$

obtaining an integral relative to the *new variable*: $= \int f(u)\, du$

Step 2. *Integrate* relative to the *new variable* u: $= F(u) + C$

Step 3. *Substitute back* returning to the *old variable* x: $= F(g(x)) + C$.

3.1.2.   *Perfect Substitution*

When the *Substitution Method* can be applied directly, without manipulating the *integrand*, we have the case of a *perfect substitution*.

Examples 3.1 (Perfect Substitution).

1. $\displaystyle\int 10^{\sin x} \cos x\, dx$ *substituting:* $\boxed{u = \sin x,\ \ du = \cos x\, dx}$;

$\displaystyle= \int 10^u\, du$ integrating relative to the new variable;

$$= \frac{10^u}{\ln 10} + C \qquad\qquad\qquad\qquad\qquad \textit{substituting back;}$$

$$= \frac{10^{\sin x}}{\ln 10} + C.$$

2. $\displaystyle\int \cot x \, dx$ $\qquad$ by the *trigonometric identity,* $\quad \boxed{\cot x = \dfrac{\cos x}{\sin x}}$;

$$= \int \frac{\cos x}{\sin x} \, dx \qquad\qquad \textit{substituting:} \quad \boxed{u = \sin x, \ du = \cos x \, dx} ;$$

$$= \int \frac{1}{u} \, du \qquad\qquad\quad \text{integrating relative to the new variable;}$$

$$= \ln |u| + C \qquad\qquad\qquad\qquad\qquad\qquad \textit{substituting back;}$$

$$= \ln |\sin x| + C \qquad (\textit{Table of Basic Integrals} \ (\text{Appendix A}),\ \text{integral 12}).$$

3. $\displaystyle\int \frac{e^x}{e^{2x} + 1} \, dx = \int \frac{1}{\left(e^x\right)^2 + 1} e^x \, dx \ \ \textit{substituting:} \ \ \boxed{u = e^x, \ u = e^x \, dx} ;$

$$= \int \frac{1}{u^2 + 1} \, du \qquad\qquad \text{integrating relative to the new variable;}$$

$$= \arctan u + C \qquad\qquad\qquad\qquad\qquad \textit{substituting back;}$$

$$= \arctan e^x + C.$$

4. $\displaystyle\int \frac{\ln x}{x} \, dx \qquad\qquad\qquad \textit{substituting:} \quad \boxed{u = \ln x, \ du = \dfrac{1}{x} \, dx} ;$

$$= \int u \, du \qquad\qquad\qquad \text{integrating relative to the new variable;}$$

$$= \frac{u^2}{2} + C \qquad\qquad\qquad\qquad\qquad\qquad \textit{substituting back;}$$

$$= \frac{\ln^2 x}{2} + C.$$

3.1.3. *Introducing a Missing Constant*

Sometimes, an integral falls short of being the case of a *perfect substitution* by a *missing constant factor* in the *integrand* only.

This can be easily fixed: the *missing constant* can be introduced via multiplying and dividing by it. After such a minor adjustment, we arrive at a perfect substitution.

Examples 3.2 (Introducing a Missing Constant).

1. $\displaystyle\int \tan x \, dx$ $\qquad\qquad$ since $\boxed{\tan x = \dfrac{\sin x}{\cos x}}$;

$\displaystyle = \int \frac{1}{\cos x} \sin x \, dx$

$\qquad$ since $\cos' x = -\sin x$, introducing the *missing constant* -1;

$\displaystyle = -\int \frac{1}{\cos x}(-\sin x)\, dx$ $\quad$ *substituting:* $\boxed{u = \cos x,\ du = -\sin x \, dx}$;

$\displaystyle = -\int \frac{1}{u}\, du$ $\qquad\qquad$ integrating relative to the new variable;

$= -\ln|u| + C$ $\qquad\qquad\qquad$ *substituting back;*

$= -\ln|\cos x| + C = \ln|\sec x| + C$

$\qquad\qquad$ (*Table of Basic Integrals* (Appendix A), integral 11).

2. $\displaystyle\int \frac{3^{1/x}}{x^2}\, dx$

$\qquad$ since $\left(\dfrac{1}{x}\right)' = -\dfrac{1}{x^2}$, introducing the *missing constant* -1;

$\displaystyle = -\int 3^{1/x}\left(-\frac{1}{x^2}\right) dx$ $\quad$ *substituting:* $\boxed{u = \dfrac{1}{x},\ du = -\dfrac{1}{x^2}\, dx}$;

$\displaystyle = -\int 3^u \, du$ $\qquad\qquad$ integrating relative to the new variable;

$\displaystyle = -\frac{3^u}{\ln 3} + C$ $\qquad\qquad\qquad$ *substituting back;*

$\displaystyle = -\frac{3^{1/x}}{\ln 3} + C.$

3. $\displaystyle\int \frac{x}{\sqrt[3]{x^2 - 1}}\, dx = \int (x^2 - 1)^{-1/3} x \, dx$

$\qquad$ since $(x^2 - 1)' = 2x$, introducing the *missing constant* 2;

$\displaystyle = \frac{1}{2}\int (x^2 - 1)^{-1/3} 2x \, dx$ $\quad$ *substituting:* $\boxed{u = x^2 - 1,\ du = 2x\, dx}$;

$$= \frac{1}{2} \int u^{-1/3} \, du \qquad \text{integrating relative to the new variable;}$$

$$= \frac{1}{2} \frac{3}{2} u^{2/3} + C = \frac{3}{4} u^{2/3} + C \qquad \text{substituting back;}$$

$$= \frac{3}{4} (x^2 - 1)^{2/3} + C.$$

4. $\int \sin(1 - 5x^4) x^3 \, dx$

since $(1 - 5x^4)' = -20x^3$, introducing the *missing constant* -20;

$$= \frac{1}{20} \int \left[-\sin(1 - 5x^4) \right] (-20x^3) \, dx$$

substituting: $\boxed{u = 1 - 5x^4, \ du = -20x^3 \, dx}$;

$$= \frac{1}{20} \int \left[-\sin u \right] du \qquad \text{integrating relative to the new variable;}$$

$$= \frac{1}{20} \cos u + C \qquad \text{substituting back;}$$

$$= \frac{1}{20} \cos(1 - 5x^4) + C.$$

5. $\int \sqrt{3x + 14} \, dx = \int (3x + 14)^{1/2} \, dx$

since $(3x + 14)' = 3$, introducing the *missing constant* 3;

$$= \frac{1}{3} \int (3x + 14)^{1/2} 3 \, dx \qquad \text{substituting: } \boxed{u = 3x + 14, \ du = 3 \, dx};$$

$$= \frac{1}{3} \int u^{1/2} \, du \qquad \text{integrating relative to the new variable;}$$

$$= \frac{1}{3} \frac{2}{3} u^{3/2} + C = \frac{2}{9} u^{3/2} + C \qquad \text{substituting back;}$$

$$= \frac{2}{9} (3x + 14)^{3/2} + C.$$

3.1.4. *Trivial Substitution*

The last problem is a typical example of an integral of the form

$$\int f(ax + b) \, dx,$$

where a, b are real coefficients with $a \neq 0$.

In such a case, a *perfect substitution* is readily attained by introducing the missing constant a:

$$\int f(ax+b)\,dx \qquad\qquad \text{introducing the } \textit{missing constant } a;$$

$$= \frac{1}{a}\int f(ax+b)a\,dx \qquad \text{substituting: } \boxed{u = ax+b,\ du = a\,dx};$$

$$= \frac{1}{a}\int f(u)\,du \qquad\qquad \text{integrating relative to the new variable;}$$

$$= \frac{1}{a}F(u) + C \qquad\qquad\qquad \text{substituting back;}$$

$$= \frac{1}{a}F(ax+b) + C.$$

We call such a routine substitution *trivial*. It is recommended to bypass a *trivial substitution* by applying the *Useful Integration Formula* instead:

$$\int f(ax+b)\,dx = \frac{1}{a}F(ax+b) + C$$

(see Theorem 2.1 and accompanying Examples 2.1).
For instance, the last problem in the preceding examples can be solved as follows:

$$\int \sqrt{3x+14}\,dx = \int (3x+14)^{1/2}\,dx = \frac{1}{3}\frac{2}{3}(3x+14)^{3/2}+C = \frac{2}{9}(3x+14)^{3/2}+C.$$

3.1.5.   *More Than a Missing Constant*

Sometimes, conducting a substitution calls for more than just noticing and introducing a missing constant factor in the integrand.

Examples 3.3 (More Than a Missing Constant).

1. $\displaystyle\int \frac{x^5}{x^3+4}\,dx \qquad\qquad\qquad\qquad \textit{splitting } x^2 \textit{ off } x^5;$

$$= \int \frac{x^3}{x^3+4}x^2\,dx$$

$$\qquad\qquad \text{since } (x^3+4)' = 3x^2, \text{ introducing the } \textit{missing constant } 3;$$

$$= \frac{1}{3}\int \frac{x^3}{x^3+4}3x^2\,dx$$

$$\qquad\qquad \text{substituting: } \boxed{u = x^3+4,\ x^3 = u-4,\ du = 3x^2\,dx};$$

$$= \frac{1}{3} \int \frac{u-4}{u}\, du \qquad\qquad\qquad \text{dividing termwise;}$$

$$= \frac{1}{3} \int \left(1 - \frac{4}{u}\right) du \qquad\qquad \text{integrating relative to the new variable;}$$

$$= \frac{1}{3} \left[\int 1\, du - 4\int \frac{1}{u}\, du\right] = \frac{1}{3}\left[u - 4\ln |u|\right] + B \qquad \text{substituting back;}$$

$$= \frac{1}{3}\left[x^3 + 4 - 4\ln |x^3 + 4|\right] + B = \frac{1}{3}x^3 - \frac{4}{3}\ln |x^3 + 4| + C \quad (C = B + 4/3).$$

2. $\displaystyle \int x^2 \sqrt[3]{x+1}\, dx = \int x^2 (x+1)^{1/3}\, dx$

$$\text{substituting:} \quad \boxed{u = x+1,\ x = u-1,\ dx = du};$$

$$= \int (u-1)^2 u^{1/3}\, du \qquad\qquad\qquad\qquad\qquad \text{multiplying;}$$

$$= \int (u^2 - 2u + 1)u^{1/3}\, du = \int \left[u^{7/3} - 2u^{4/3} + u^{1/3}\right] du$$

$$\text{integrating relative to the new variable;}$$

$$= \int u^{7/3}\, du - 2\int u^{4/3}\, du + \int u^{1/3}\, du = \frac{3}{10}u^{10/3} - \frac{6}{7}u^{7/3} + \frac{3}{4}u^{4/3} + C$$

$$\text{substituting back;}$$

$$= \frac{3}{10}(x+1)^{10/3} - \frac{6}{7}(x+1)^{7/3} + \frac{3}{4}(x+1)^{4/3} + C$$

$$= \frac{3}{140}(x+1)^{4/3}\left[14(x+1)^2 - 40(x+1) + 35\right] + C$$

$$= \frac{3}{140}(x+1)^{4/3}\left[14x^2 - 12x + 9\right] + C.$$

3. $\displaystyle \int \frac{1}{e^x - 2}\, dx \qquad\qquad\qquad \text{multiplying and dividing by } e^x;$

$$= \int \frac{e^x}{e^x(e^x - 2)}\, dx \qquad\qquad \text{substituting:} \quad \boxed{u = e^x,\ du = e^x\, dx};$$

$$= \int \frac{1}{u(u-2)}\, du \qquad\qquad\qquad \text{transforming product into sum:}$$

$$\boxed{\frac{1}{u(u-2)} = \frac{1}{2}\left[\frac{1}{u-2} - \frac{1}{u}\right]} \quad \text{(cf. Examples 2.10);}$$

$$= \int \frac{1}{2} \left[\frac{1}{u-2} - \frac{1}{u} \right] du \qquad \text{integrating relative to the new variable;}$$

$$= \frac{1}{2} \left[\int \frac{1}{u-2} \, du - \int \frac{1}{u} \, du \right] = \frac{1}{2} \left[\ln|u-2| - \ln|u| \right] + C$$

$$\qquad\qquad\qquad\qquad\qquad\qquad\qquad\qquad \textit{substituting back;}$$

$$= \frac{1}{2} \left[\ln|e^x - 2| - \ln e^x \right] + C = \frac{1}{2} \left[\ln|e^x - 2| - x \right] + C.$$

4. $\displaystyle\int \sec x \, dx$ \qquad\qquad multiplying and dividing by $\sec x + \tan x$;

$$= \int \frac{\sec x (\sec x + \tan x)}{\sec x + \tan x} \, dx = \int \frac{1}{\sec x + \tan x} (\sec x \tan x + \sec^2 x) \, dx$$

$\qquad\qquad$ *substituting:* $\boxed{u = \sec x + \tan x, \ du = (\sec x \tan x + \sec^2 x) \, dx}$;

$$= \int \frac{1}{u} \, du \qquad\qquad \text{integrating relative to the new variable;}$$

$$= \ln|u| + C \qquad\qquad\qquad\qquad\qquad\qquad \textit{substituting back;}$$

$$= \ln|\sec x + \tan x| + C$$

$$\qquad\qquad\qquad (\textit{Table of Basic Integrals} \text{ (Appendix A), integral 13).}$$

Exercise 3.1 (Deriving Table Integrals).
In the similar manner, obtain integral 14 of the *Table of Basic Integrals* (Appendix A).

3.1.6. *More Than One Way*

There may exist multiple approaches to finding an indefinite integral, some of which involve substitution as in the following case.

Example 3.4 (More Than One Way).
Evaluate the integral

$$\int x\sqrt{5 - 2x} \, dx.$$

Approach 1 (Direct Integration): This approach steers clear of using substitution. A complete solution is given in Examples 2.5.

Approach 2 (Substitution):

$$\int x\sqrt{5-2x}\,dx \qquad \text{substituting:} \quad \boxed{u = 5-2x,\ x = \frac{5-u}{2},\ dx = -\frac{1}{2}du}\ ;$$

$$= \int \frac{5-u}{2}u^{1/2}\left(-\frac{1}{2}\right)du \qquad\qquad \text{multiplying;}$$

$$= \int \frac{1}{4}(u-5)u^{1/2}\,du = \int \frac{1}{4}(u^{3/2} - 5u^{1/2})\,du$$

$$\text{integrating relative to the new variable;}$$

$$= \frac{1}{4}\int (u^{3/2} - 5u^{1/2})\,du = \frac{1}{4}\left[\int u^{3/2}\,du - 5\int u^{1/2}\,du\right]$$

$$= \frac{1}{4}\left[\frac{2}{5}u^{5/2} - \frac{10}{3}u^{3/2}\right] + C \qquad\qquad \text{substituting back;}$$

$$= \frac{1}{10}(5-2x)^{5/2} - \frac{5}{6}(5-2x)^{3/2} + C.$$

Approach 3 (Another Substitution): Let us use the substitution $u = \sqrt{5-2x}$ now.

$$\int x\sqrt{5-2x}\,dx$$

$$\text{substituting:} \quad \boxed{u = \sqrt{5-2x},\ x = \frac{5-u^2}{2},\ dx = -u\,du}\ ;$$

$$= \int \frac{5-u^2}{2}u(-u)\,du \qquad\qquad \text{multiplying;}$$

$$= \int \frac{1}{2}(u^4 - 5u^2)\,du \qquad\quad \text{integrating relative to the new variable;}$$

$$= \frac{1}{2}\left[\int u^4\,du - 5\int u^2\,du\right] = \frac{1}{2}\left[\frac{1}{5}u^5 - \frac{5}{3}u^3\right] + C \qquad \text{substituting back;}$$

$$= \frac{1}{10}(5-2x)^{5/2} - \frac{5}{6}(5-2x)^{3/2} + C.$$

3.1.7.   *More Than One Substitution*

Substitutions may be applied repeatedly.

Examples 3.5 (More Than One Substitution).

1. $\int x\sin^3(x^2)\cos(x^2)\,dx \qquad\qquad \text{introducing the *missing constant* 2;}$

$$= \frac{1}{2} \int \sin^3(x^2) \cos(x^2) 2x \, dx \qquad \text{substituting:} \quad \boxed{u = x^2, \ du = 2x \, dx};$$

$$= \frac{1}{2} \int \sin^3 u \cos u \, du \qquad \text{substituting again:} \quad \boxed{v = \sin u, \ dv = \cos u \, du};$$

$$= \frac{1}{2} \int v^3 \, dv \qquad \qquad \text{integrating relative to the new variable;}$$

$$= \frac{1}{8} v^4 + C \qquad \qquad \text{substituting back;}$$

$$= \frac{1}{8} \sin^4 u + C = \frac{1}{8} \sin^4(x^2) + C$$

2. $\displaystyle \int \frac{\sin^3 x \cos x}{\sin^2 x + 1} \, dx$ $\qquad$ substituting: $\boxed{u = \sin x, \ du = \cos x \, dx}$;

$$= \int \frac{u^3}{u^2 + 1} \, du \qquad \text{by } long \ division, \quad \boxed{\frac{u^3}{u^2 + 1} = u - \frac{u}{u^2 + 1}} \quad (verify);$$

$$= \int u \, du - \int \frac{u}{u^2 + 1} \, du \qquad \text{introducing the } missing \ constant \ 2;$$

$$= \frac{u^2}{2} - \frac{1}{2} \int \frac{2u}{u^2 + 1} \, du \qquad \text{substituting again:} \quad \boxed{v = u^2 + 1, \ dv = 2u \, du};$$

$$= \frac{u^2}{2} - \frac{1}{2} \int \frac{1}{v} \, dv \qquad \text{integrating relative to the new variable;}$$

$$= \frac{u^2}{2} - \frac{1}{2} \ln|v| + C \qquad \text{substituting back;}$$

$$= \frac{\sin^2 x}{2} - \frac{1}{2} \ln(\sin^2 x + 1) + C.$$

Remark 3.2. When two or more consecutive substitutions can be applied, a single substitution, perhaps a bulkier one, can always be applied instead.

Thus, the second integral in the prior examples can also be evaluated via a single substitution as follows:

Example 3.6 (Last Integral Revisited).

$$\int \frac{\sin^3 x \cos x}{\sin^2 x + 1} \, dx \qquad \text{splitting } \sin x \text{ off } \sin^3 x;$$

$$= \int \frac{\sin^2 x}{\sin^2 x + 1} \sin x \cos x \, dx \qquad \text{introducing the } missing \ constant \ 2;$$

$$= \frac{1}{2} \int \frac{\sin^2 x}{\sin^2 x + 1} 2 \sin x \cos x \, dx$$

substituting: $\boxed{u = \sin^2 x + 1, \ \sin^2 x = u - 1, \ du = 2\sin x \cos x \, dx}$;

$$= \frac{1}{2} \int \frac{u-1}{u} \, du \qquad\qquad \text{dividing termwise;}$$

$$= \frac{1}{2} \int \left(1 - \frac{1}{u}\right) du \qquad\qquad \text{integrating relative to the new variable;}$$

$$= \frac{1}{2} \left[\int 1 \, du - \int \frac{1}{u} \, du\right] = \frac{1}{2} \left[u - \ln|u|\right] + C \qquad\qquad \text{substituting back;}$$

$$= \frac{1}{2} \left[\sin^2 x + 1 - \ln(\sin^2 x + 1)\right] + C = \frac{1}{2} \sin^2 x - \frac{1}{2} \ln(\sin^2 x + 1) + B$$

$$(B = C + 1/2).$$

Exercise 3.2 (Alternate Approach).
Evaluate the first integral in the prior examples via the single substitution $u = \sin(x^2)$.

3.2. Substitution for Definite Integral

3.2.1. *What for? Why? How?*

What Is the Method for?
The *Method of Substitution* for definite integral is used for finding integrals of the form:

$$\int_a^b f(g(x))g'(x) \, dx,$$

e.g.,

$$\int_1^e \frac{\sqrt{\ln x + 1}}{x} \, dx$$

with $f(u) = \sqrt{u}$ and $g(x) = \ln x + 1$.

Why Does the Method Work?
The method is based on the following

Theorem 3.2 (Substitution Rule for Definite Integral).
Let

- *a function f be continuous on an interval I and*

- *a function g be continuously differentiable on an interval $[a, b]$ and take values in I.*

Then, with the substitution $u = g(x)$,

$$\int_a^b f(g(x))g'(x)\,dx = \int_{g(a)}^{g(b)} f(u)\,du.$$

Proof. Observe that the integrals on both sides of the equality exist since the integrands are functions *continuous* on the corresponding intervals.

Since f is continuous on I, by the *Fundamental Theorem of Calculus (Part 1)* (Theorem 1.5), it has an antiderivative $F(x)$ on I.

By the *Substitution Rule for Indefinite Integral* (Theorem 3.1),

$$\int f(g(x))g'(x)\,dx = F(g(x)) + C.$$

Hence, by the *Newton-Leibniz Formula* (Theorem 1.6),

$$\int_a^b f(g(x))g'(x)\,dx = F(g(x))\Big|_a^b = F(g(b)) - F(g(a)) = F(u)\Big|_{g(a)}^{g(b)}$$
$$= \int_{g(a)}^{g(b)} f(u)\,du.$$

$\square$

How Does the Method Work?

An integral of the form

$$\int_a^b f(g(x))g'(x)\,dx,$$

with functions $f(x)$ and $g(x)$ satisfying the conditions of the *Substitution Rule*, is found in the following *three steps*:

$$\int_a^b f(g(x))g'(x)\,dx$$

Step 1. *Substitute* and *change the integration limits*

$u = g(x), \ du = g'(x)\,dx$	x	u
	a	$g(a)$
	b	$g(b)$

:

$$= \int_{g(a)}^{g(b)} f(u)\, du$$

Step 2. *Integrate (antidifferentiate)* relative to the *new variable u.*

Step 3. Use the *Newton-Leibniz Formula*:

$$= F(u)\Big|_{g(a)}^{g(b)} = F(g(b)) - F(g(a)).$$

Remark 3.3. The *Substitution Method* for definite integral

- does not require *back substitution*, but
- requires the *change* of the *integration limits*.

Basically, when executing substitution for a definite integral, we act as if we have an indefinite integral, not forgetting to *change* the *integration limits*, up to the step of *back substitution*, instead of which, we find the numeric value of the obtained definite integral relative to the new integration variable and new integration limits by the *Newton-Leibniz Formula*.

Examples 3.7 (Substitution for Definite Integral).

1. $\displaystyle\int_1^e \frac{\sqrt{\ln x + 1}}{x}\, dx = \int_1^e (\ln x + 1)^{1/2}\frac{1}{x}\, dx$

substituting and *changing* the *integration limits*:

$$u = \ln x + 1,\ du = \frac{1}{x}\, dx \quad \begin{array}{c|c} x & u \\ \hline 1 & 1 \\ e & 2 \end{array};$$

$$= \int_1^2 u^{1/2}\, du \qquad \text{by the *Newton-Leibniz Formula*;}$$

$$= \frac{2}{3}u^{3/2}\Big|_1^2 = \frac{2}{3}[2^{3/2} - 1] = \frac{4}{3}\sqrt{2} - \frac{2}{3}.$$

2. $\displaystyle\int_0^1 xe^{-x^2}\, dx \qquad \text{introducing the *missing constant* } -2;$

$$= -\frac{1}{2}\int_0^1 e^{-x^2}(-2x)\, dx$$

substituting and *changing* the *integration limits*:

$$
u = -x^2, \; du = -2x\,dx \quad
\begin{array}{c|c}
x & u \\
\hline
0 & 0 \\
1 & -1
\end{array} \; ;
$$

$$
= -\frac{1}{2} \int_0^{-1} e^u \, du \qquad\qquad \text{by the } \textit{Newton-Leibniz Formula};
$$

$$
= -\frac{1}{2} e^u \Big|_0^{-1} = -\frac{1}{2}(e^{-1} - e^0) = \frac{1}{2}\left(1 - \frac{1}{e}\right).
$$

3.3.   Applications

Example 3.8 (Application).
Find the *volume* of the solid generated by revolving the region bounded by $y = \sqrt{4 - x^2}$ and the x-axis on the interval $[1, 2]$ about the y-axis.

Solution: Since the rotation axis is perpendicular to the axis of definition, by the *shell method*,

$$
V = \int_1^2 2\pi x \sqrt{4 - x^2} \, dx
$$

factoring out π and switching to the *power form*;

$$
= \pi \int_1^2 (4 - x^2)^{1/2} 2x \, dx \qquad \text{introducing the } \textit{missing constant} \; -1;
$$

$$
= -\pi \int_1^2 (4 - x^2)^{1/2}(-2x) \, dx
$$

substituting and *changing* the *integration limits*:

$$
u = 4 - x^2, \; du = (-2x)\,dx \quad
\begin{array}{c|c}
x & u \\
\hline
1 & 3 \\
2 & 0
\end{array} \; ;
$$

$$
= -\pi \int_3^0 u^{1/2} \, du \qquad\qquad \text{by the } \textit{Newton-Leibniz Formula};
$$

$$
= -\pi \frac{2}{3} u^{3/2} \Big|_3^0 = -\pi \frac{2}{3}[0^{3/2} - 3^{3/2}] = 2\pi\sqrt{3} \text{ un.}^3.
$$

3.4. Practice Problems

Evaluate the integrals.

1. $\int \dfrac{2x-3}{\sqrt{x^2-3x+4}}\,dx$

2. $\int \dfrac{e^x}{\sqrt{e^{2x}-1}}\,dx$

3. $\int \dfrac{\arcsin x}{\sqrt{1-x^2}}\,dx$

4. $\int \dfrac{1}{x(\ln x-3)}\,dx$

5. $\int \dfrac{1}{x\ln x\ln|\ln x|}\,dx$

6. $\int \dfrac{x}{x^2+2}\,dx$

7. $\int \dfrac{3x+2}{x^2-5}\,dx$

8. $\int \sin x\cos x\,dx$

9. $\int \dfrac{\sin x}{\cos^3 x}\,dx$

10. $\int \dfrac{\sin x+\cos x}{\sqrt[3]{\sin x-\cos x}}\,dx$

11. $\int \dfrac{\sin x}{\sqrt{\cos 2x}}\,dx$

12. $\int e^{x^3}x^2\,dx$

13. $\int \dfrac{5^{\sqrt{x}}}{\sqrt{x}}\,dx$

14. $\int \sqrt[3]{1-2x^6}\,x^5\,dx$

15. $\int x^3(1-3x^2)^{10}\,dx$

16. $\int \dfrac{x^2}{x^6+4}\,dx$

17. $\int \dfrac{x^3}{\sqrt{x^8-1}}\,dx$

18. $\int \dfrac{x}{x^2+x+1}\,dx$

19. $\int \dfrac{x-1}{\sqrt{1-x^2}}\,dx$

20. $\int \dfrac{x^5}{\sqrt{1-x^2}}\,dx$

21. $\int \dfrac{1}{x\sqrt{x^2+1}}\,dx$

22. $\int \dfrac{1}{\sin^2 x+2\cos^2 x}\,dx$

23. $\int \dfrac{\sin(1/x)}{x^2}\,dx$

24. $\int \dfrac{1}{\sqrt{e^x+1}}\,dx$

25. $\int_0^1 \dfrac{\arctan x}{x^2+1}\,dx$

26. $\int_0^{\ln 2} \dfrac{1}{e^x+1}\,dx$

Chapter 4

Method of Integration by Parts

4.1. Partial Integration for Indefinite Integral

4.1.1. *What for? Why? How?*

What Is the Method for?

The *Method of Integration by Parts* (or *Partial Integration*) for indefinite integral is used for finding integrals of the form:

$$\int u(x)v'(x)\,dx,$$

where u and v are *continuously differentiable* functions (i.e., functions with continuous derivatives) on an interval I, e.g.,

$$\int e^x \cos x\,dx$$

with $u(x) = e^x$ and $v(x) = \sin x$.

Why Does the Method Work?

The method is based on the following

Theorem 4.1 (Integration by Parts Formula for Indefinite Integral).

Let u and v be continuously differentiable functions on an interval I. Then, on I,

$$\int u(x)v'(x)\,dx = u(x)v(x) - \int v(x)u'(x)\,dx.$$

Proof. Since the functions u and v are *continuously differentiable* on I, then so is their product uv and, by the differentiation *product rule* (see, e.g., [1, 6]),

$$[u(x)v(x)]' = u'(x)v(x) + u(x)v'(x) \quad \text{on } I,$$

which implies that uv is an *antiderivative* of the function $u'v + uv'$ on I. Thus,

$$\int \left[u'(x)v(x) + u(x)v'(x) \right] dx = u(x)v(x) + C \quad \text{on } I.$$

Since the functions $u'v$ and uv' are *continuous* on I, by the *Fundamental Theorem of Calculus (Part 1)* (Theorem 1.5), they are integrable on I and, by the *integration rules* (Theorem 1.1),

$$\int u'(x)v(x)\, dx + \int u(x)v'(x)\, dx = u(x)v(x) + C \quad \text{on } I.$$

Whence, the partial integration formula follows immediately.   □

Remarks 4.1 (Partial Integration Formula).

- Thus, the *Partial Integration Formula* is the integral form of the *Product Rule* for differentiation.
- Using the differential notation $dv = v'(x)dx$, $du = u'(x)dx$, we can informally rewrite the partial integration formula in the following easy-to-remember shorthand fashion:

$$\int u\, dv = uv - \int v\, du.$$

How Does the Method Work?

Partial integration for indefinite integral consists in applying the *integrations by parts formula* and works nicely for the following three special types of integrals (and not only for them).

4.1.2. *Three Special Types of Integrals*

Type 1 Integrals

Type 1 integrals are of the form

$$\int P_n(x)e^{ax+b}\, dx, \quad \int P_n(x)\sin(ax+b)\, dx, \quad \text{or} \quad \int P_n(x)\cos(ax+b)\, dx$$

where

$$P_n(x) = a_n x^n + a_{n-1}x^{n-1} + \cdots + a_1 x + a_0$$

is a *polynomial* of degree $n = 1, 2, \ldots$ with real coefficients $a_n, \ldots, a_0$, $a_n \neq 0$, a and b are real coefficients, $a \neq 0$.

To such integrals, we apply partial integration *precisely n times* setting on each step

$$u = polynomial\ factor \quad \text{and} \quad dv = [exponential/trigonometric\ factor]\ dx.$$

Example 4.1 (Type 1 Integrals).

$$\int (x^2 + 3x)e^{2x}\ dx$$

integrating by parts:
$$\boxed{\begin{array}{l} u = x^2 + 3x,\ du = (2x + 3)\ dx \\ dv = e^{2x}\ dx,\ v = \int e^{2x}\ dx = \dfrac{1}{2}e^{2x} \end{array}};$$

$$= (x^2 + 3x)\frac{1}{2}e^{2x} - \frac{1}{2}\int (2x + 3)e^{2x}\ dx$$

integrating by parts once more:
$$\boxed{\begin{array}{l} u = 2x + 3,\ \ du = 2\ dx \\ dv = e^{2x}\ dx,\ v = \dfrac{1}{2}e^{2x} \end{array}};$$

$$= \frac{1}{2}(x^2 + 3x)e^{2x} - \frac{1}{2}\left[\frac{1}{2}(2x + 3)e^{2x} - \int e^{2x}\ dx\right]$$

$$= \frac{1}{2}(x^2 + 3x)e^{2x} - \frac{1}{2}\left[\frac{1}{2}(2x + 3)e^{2x} - \frac{1}{2}e^{2x}\right] + C \qquad \text{simplifying;}$$

$$= \frac{1}{2}\left[x^2 + 3x - \frac{1}{2}(2x + 3) + \frac{1}{2}\right]e^{2x} + C = \frac{1}{2}\left(x^2 + 2x - 1\right)e^{2x} + C.$$

Type 2 Integrals

Type 2 integrals are of the form

$$\int P_n(x)\ln(ax + b)\ dx,$$

$$\int P_n(x)\arcsin(ax + b)\ dx, \quad \int P_n(x)\arccos(ax + b)\ dx,$$

$$\int P_n(x)\arctan(ax + b)\ dx, \quad \text{or} \quad \int P_n(x)\text{arccot}(ax + b)\ dx,$$

where

$$P_n(x) = a_n x^n + a_{n-1}x^{n-1} + \cdots + a_1 x + a_0$$

is a *polynomial* of degree $n = 0, 1, 2, \ldots$ with real coefficients $a_n, \ldots, a_0$, $a_n \neq 0$, a and b are real coefficients, $a \neq 0$.

To such integrals, we apply partial integration *once* setting

$$u = transcendental\ factor \quad and \quad dv = P_n(x)\,dx.$$

Example 4.2 (Type 2 Integrals).

$$\int x \ln x\,dx \qquad integrating\ by\ parts: \qquad \boxed{\begin{array}{l} u = \ln x, \quad du = \dfrac{1}{x}\,dx \\[2mm] dv = x\,dx,\ v = \int x\,dx = \dfrac{x^2}{2} \end{array}};$$

$$= \frac{x^2}{2}\ln x - \frac{1}{2}\int x^2\frac{1}{x}\,dx = \frac{x^2}{2}\ln x - \frac{1}{2}\int x\,dx = \frac{x^2}{2}\ln x - \frac{x^2}{4} + C$$

$$= \frac{x^2}{4}(2\ln x - 1) + C.$$

Type 3 Integrals

Type 3 integrals are of the form

$$\int e^{ax+b}\sin(cx+d)\,dx \quad or \quad \int e^{ax+b}\cos(cx+d)\,dx$$

where a, b, c, and d are real coefficients, $a, c \neq 0$.

To such integrals, we apply partial integration *twice* setting on each step

$$u = exponential\ factor \quad and \quad dv = [trigonometric\ factor]\,dx$$

or vice versa and arrive at a *linear algebraic equation* relative to the unknown integral, solving which we find the integral.

Example 4.3 (Type 3 Integrals).

$$\int e^{-x}\sin x\,dx$$

$$integrating\ by\ parts: \qquad \boxed{\begin{array}{l} u = e^{-x}, \quad\quad du = -e^{-x}\,dx \\[1mm] dv = \sin x\,dx,\ v = \int \sin x\,dx = -\cos x \end{array}};$$

$$= -e^{-x}\cos x - \int e^{-x}\cos x\,dx$$

$$integrating\ by\ parts\ once\ more: \qquad \boxed{\begin{array}{l} u = e^{-x}, \quad\quad du = -e^{-x}\,dx \\[1mm] dv = \cos x\,dx,\ v = \int \cos x\,dx = \sin x \end{array}};$$

$$= -e^{-x}\cos x - \left[e^{-x}\sin x + \int e^{-x}\sin x\,dx \right].$$

Thus, we arrive at the following equation for the unknown integral:

$$\int e^{-x} \sin x \, dx = -e^{-x}(\cos x + \sin x) - \int e^{-x} \sin x \, dx,$$

solving which, we obtain

$$\int e^{-x} \sin x \, dx = -\frac{1}{2}e^{-x}(\cos x + \sin x) + C.$$

4.1.3. *Beyond Three Special Types*

As the following examples demonstrate, the applicability of partial integration extends beyond the three considered types of integrals.

Examples 4.4 (Beyond Three Special Types).

1. $\displaystyle\int \ln(x^2 + 1)\, dx$ applying *type* 2 partial integration scenario:

$$u = \ln(x^2 + 1), \ du = \frac{2x}{x^2 + 1}\, dx \ ;$$
$$dv = dx, \qquad v = x$$

$$= x \ln(x^2 + 1) - 2\int \frac{x^2}{x^2 + 1}\, dx \qquad \text{integrating directly;}$$

$$= x \ln(x^2 + 1) - 2\int \frac{x^2 + 1 - 1}{x^2 + 1}\, dx = x \ln(x^2 + 1) - 2\int \left[1 - \frac{1}{x^2 + 1}\right] dx$$

$$= x \ln(x^2 + 1) - 2\left[\int 1\, dx - \int \frac{1}{x^2 + 1}\, dx\right]$$

$$= x \ln(x^2 + 1) - 2\left[x - \arctan x\right] + C.$$

2. $\displaystyle\int \frac{\arcsin x}{\sqrt{1 + x}}\, dx$ applying *type* 2 partial integration scenario:

$$u = \arcsin x, \qquad du = \frac{1}{\sqrt{1 - x^2}}\, dx$$
$$dv = \frac{1}{\sqrt{1 + x}}\, dx, \ v = \int (1 + x)^{-1/2}\, dx = 2(1 + x)^{1/2} \ ;$$

$$= 2\sqrt{1 + x}\, \arcsin x - 2\int \frac{\sqrt{1 + x}}{\sqrt{1 - x^2}}\, dx = 2\sqrt{1 + x}\, \arcsin x - 2\int \frac{1}{\sqrt{1 - x}}\, dx$$

$$= 2\sqrt{1 + x}\, \arcsin x - 2\int (1 - x)^{-1/2}\, dx = 2\sqrt{1 + x}\, \arcsin x + 4\sqrt{1 - x} + C.$$

3. $\displaystyle\int x \tan^2 x\, dx$

using the relevant *trigonometric identity*: $\boxed{\tan^2 x = \sec^2 x - 1}$;

$$= \int x(\sec^2 x - 1)\, dx = \int x \sec^2 x\, dx - \int x\, dx$$

integrating by parts: $\boxed{\begin{array}{ll} u = x, & du = dx \\ dv = \sec^2 x\, dx, & v = \int \sec^2 x\, dx = \tan x \end{array}}$;

$$= x \tan x - \int \tan x\, dx - \frac{x^2}{2} = x \tan x - \ln|\sec x| - \frac{x^2}{2} + C.$$

4. For $a > 0$,

$$\int \sqrt{x^2 \pm a^2}\, dx$$

integrating by parts: $\boxed{\begin{array}{ll} u = \sqrt{x^2 \pm a^2}, & du = \dfrac{x}{\sqrt{x^2 \pm a^2}}\, dx \\ dv = dx, & v = x \end{array}}$;

$$= x\sqrt{x^2 \pm a^2} - \int \frac{x^2}{\sqrt{x^2 \pm a^2}}\, dx$$

adding/subtracting and subtracting/adding a^2;

$$= x\sqrt{x^2 \pm a^2} - \int \frac{x^2 \pm a^2 \mp a^2}{\sqrt{x^2 \pm a^2}}\, dx \qquad\qquad \text{dividing termwise;}$$

$$= x\sqrt{x^2 \pm a^2} - \int \left[\sqrt{x^2 \pm a^2} \mp \frac{a^2}{\sqrt{x^2 \pm a^2}} \right] dx$$

by the *integration rules*;

$$= x\sqrt{x^2 \pm a^2} - \int \sqrt{x^2 \pm a^2}\, dx \pm a^2 \int \frac{1}{\sqrt{x^2 \pm a^2}}\, dx$$

$$= x\sqrt{x^2 \pm a^2} - \int \sqrt{x^2 \pm a^2}\, dx \pm a^2 \ln|x + \sqrt{x^2 \pm a^2}|.$$

Thus, we arrive at the following equation for the unknown integral:

$$\int \sqrt{x^2 \pm a^2}\, dx = x\sqrt{x^2 \pm a^2} - \int \sqrt{x^2 \pm a^2}\, dx \pm a^2 \ln|x + \sqrt{x^2 \pm a^2}|,$$

solving which, we obtain

$$\int \sqrt{x^2 \pm a^2}\, dx = \frac{x}{2}\sqrt{x^2 \pm a^2} \pm \frac{a^2}{2} \ln|x + \sqrt{x^2 \pm a^2}| + C.$$

4.1.4. *Reduction Formulas*

When integrating

$$\int x^3 e^x\, dx,$$

we apply *partial integration* repeatedly, progressively reducing the power of x until we arrive at the easy-to-evaluate integral

$$\int e^x\, dx = e^x + C.$$

Each step in this process is executed under the same pattern, which can be described by a corresponding *reduction formula*, i.e., an integral identity, in which the power in the integrand is reduced. When applied repeatedly, such formula progressively simplifies the integral until it can be evaluated directly.

Reduction formulas become really handy when dealing with integrals containing a high power. Some of them can be found in Appendix B. Let us establish a few here.

Example 4.5 (Reduction Formulas).

Appendix B, formula 8. For $a \neq 0$ and $n = 1, 2, \ldots$,

$$\int x^n e^{ax}\, dx \qquad\qquad \text{applying } type \text{ 1 partial integrating scenario:}$$

$$\boxed{\begin{aligned} u &= x^n, \qquad du = nx^{n-1}\, dx \\ dv &= e^{ax}\, dx,\ v = \int e^{ax}\, dx = \frac{1}{a}e^{ax} \end{aligned}};$$

$$= \frac{x^n e^{ax}}{a} - \frac{n}{a}\int x^{n-1} e^{ax}\, dx.$$

Applying the obtained reduction formula *threefold* to the integral

$$\int x^3 e^x\, dx$$

($a = 1$ and $n = 3$), we have:

$$\int x^3 e^x \, dx = x^3 e^x - 3 \int x^2 e^x \, dx = x^3 e^x - 3 \left[x^2 e^x - 2 \int x e^x \, dx \right]$$

$$= x^3 e^x - 3x^2 e^x + 6 \left[x e^x - \int e^x \, dx \right] = \left[x^3 - 3x^2 + 6x - 6 \right] e^x + C.$$

Exercise 4.1 (Reduction Formulas).
Prove reduction formulas 9 and 10 of Appendix B.

Examples 4.6 (Reduction Formulas).

1. Appendix B, formula 7. For $n = 1, 2, \ldots$,

$$\int \ln^n x \, dx$$

applying *type* 2 partial integrating scenario:

$$\boxed{\begin{aligned} u &= \ln^n x, \ du = n \ln^{n-1} x \frac{1}{x} \, dx \\ dv &= dx, \ \ v = x \end{aligned}};$$

$$= x \ln^n x - n \int \ln^{n-1} x \, dx.$$

2. Appendix B, formula 11. For $a > 0$ and $n = 2, 3, \ldots$,

$$\int \frac{1}{(x^2 + a^2)^{n-1}} \, dx \hspace{3cm} \text{\textit{integrating by parts:}}$$

$$\boxed{\begin{aligned} u &= \frac{1}{(x^2 + a^2)^{n-1}} = (x^2 + a^2)^{-(n-1)}, \ du = -2(n-1) \frac{x}{(x^2 + a^2)^n} \, dx \\ dv &= dx, \hspace{4cm} v = x \end{aligned}};$$

$$= \frac{x}{(x^2 + a^2)^{n-1}} + (2n - 2) \int \frac{x^2}{(x^2 + a^2)^n} \, dx$$

adding and subtracting a^2 in the numerator;

$$= \frac{x}{(x^2 + a^2)^{n-1}} + (2n - 2) \int \frac{x^2 + a^2 - a^2}{(x^2 + a^2)^n} \, dx$$

$$= \frac{x}{(x^2 + a^2)^{n-1}} + (2n - 2) \left[\int \frac{1}{(x^2 + a^2)^{n-1}} \, dx - a^2 \int \frac{1}{(x^2 + a^2)^n} \, dx \right].$$

Thus, we arrive at the following equation for $\int \frac{1}{(x^2 + a^2)^n} \, dx$:

$$\int \frac{1}{(x^2+a^2)^{n-1}} \, dx = \frac{x}{(x^2+a^2)^{n-1}} + (2n-2) \int \frac{1}{(x^2+a^2)^{n-1}} \, dx$$
$$- (2n-2)a^2 \int \frac{1}{(x^2+a^2)^n} \, dx,$$

solving which, we obtain formula 11 of Appendix B:

$$\int \frac{1}{(x^2+a^2)^n} \, dx$$
$$= \frac{1}{(2n-2)a^2} \left[\frac{x}{(x^2+a^2)^{n-1}} + (2n-3) \int \frac{1}{(x^2+a^2)^{n-1}} \, dx \right] + C.$$

Remark 4.2. The latter formula is useful for the integration of *rational functions* (see Sec. 7.2.3). Integration via reduction formulas is also utilized for trigonometric integrals (see Sec. 5.2.1).

4.2. Partial Integration for Definite Integral

4.2.1. *What for? Why? How?*

What Is the Method for?

The *Method of Integration by Parts* (or *Partial Integration*) for definite integral is used for evaluation of integrals of the form:

$$\int_a^b u(x)v'(x) \, dx,$$

where u and v are continuously differentiable functions on an interval $[a, b]$ $(-\infty < a < b < \infty)$, e.g.,

$$\int_0^\pi x^2 e^x \, dx$$

with $u(x) = x^2$ and $v(x) = e^x$.

Why Does the Method Work?

The method is based on the following

Theorem 4.2 (Integration by Parts Formula for Definite Integral).
Let u and v be continuously differentiable functions on an interval $[a, b]$ $(-\infty < a < b < \infty)$. Then

$$\int_a^b u(x)v'(x) \, dx = u(x)v(x) \Big|_a^b - \int_a^b v(x)u'(x) \, dx.$$

Proof. Since the functions u and v are continuously differentiable on $[a, b]$, then so is their product uv and, by the differentiation *Product Rule* (see, e.g., [1, 6]), uv is an *antiderivative* of the function $u'v + uv'$ on $[a, b]$ (cf. the proof of Theorem 4.1).

Hence, by the *Newton-Leibniz Formula* (Theorem 1.6), we have:

$$\int_a^b \left[u'(x)v(x) + u(x)v'(x) \right] dx = u(x)v(x) \Big|_a^b.$$

Since the functions $u'v$ and uv' are *continuous* on $[a, b]$, they are *(Riemann-) integrable* on $[a, b]$ (Corollary 1.1) and, by the *integration rules* (Theorem 1.3),

$$\int_a^b u'(x)v(x)\, dx + \int_a^b u(x)v'(x)\, dx = u(x)v(x) \Big|_a^b.$$

Whence, the partial integration formula follows immediately. $\square$

Remark 4.3. Informally, the partial integration formula can be written as follows:

$$\int_a^b u\, dv = uv \Big|_a^b - \int_a^b v\, du.$$

How Does the Method Work?

Partial integration for definite integral, as well as for indefinite one, consists in applying the formula of integrations by parts.

Partial integration works for definite integral whenever and in the same manner it does for indefinite integral, in particular, for the three special types of integrals.

Remark 4.4. When applying *partial integration* to a definite integral, as opposed to the *substitution method*, we *do not* change the integration limits.

Examples 4.7 (Partial Integration for Definite Integral).

1. $\displaystyle\int_0^\pi (x - \pi) \sin(7x + \pi)\, dx$

applying *type* 1 partial integration scenario:

$$\begin{array}{ll} u = x - \pi, & du = dx \\ dv = \sin(7x + \pi)\, dx, & v = \int \sin(7x + \pi)\, dx = -\dfrac{1}{7}\cos(7x + \pi) \end{array};$$

$$= -\frac{1}{7}(x - \pi)\cos(7x + \pi)\Big|_0^\pi + \frac{1}{7}\int_0^\pi \cos(7x + \pi)\,dx$$

$$= -\frac{1}{7}(x - \pi)\cos(7x + \pi)\Big|_0^\pi + \frac{1}{49}\sin(7x + \pi)\Big|_0^\pi = -\frac{1}{7}\pi\cos\pi$$

$$+ \frac{1}{49}(\sin 8\pi - \sin\pi) = \frac{\pi}{7}.$$

2. $\displaystyle\int_0^1 x\arctan x\,dx$ applying *type 2* partial integration scenario:

$$\boxed{\begin{array}{l} u = \arctan x, \ du = \dfrac{1}{x^2 + 1}\,dx \\[2mm] dv = x\,dx, \qquad v = \int x\,dx = \dfrac{x^2}{2} \end{array}};$$

$$= \frac{x^2 \arctan x}{2}\Big|_0^1 - \frac{1}{2}\int_0^1 \frac{x^2}{x^2 + 1}\,dx \qquad\qquad \text{integrating directly;}$$

$$= \frac{\arctan 1}{2} - \frac{1}{2}\int_0^1 \frac{x^2 + 1 - 1}{x^2 + 1}\,dx = \frac{\pi}{8} - \frac{1}{2}\int_0^1\left[1 - \frac{1}{x^2 + 1}\right]dx$$

$$= \frac{\pi}{8} - \frac{1}{2}\left[\int_0^1 1\,dx - \int_0^1 \frac{1}{x^2 + 1}\,dx\right] = \frac{\pi}{8} - \frac{1}{2}\left[x\Big|_0^1 - \arctan x\Big|_0^1\right]$$

$$= \frac{\pi}{8} - \frac{1}{2}\left[1 - \arctan 1 + \arctan 0\right] = \frac{\pi}{8} - \frac{1}{2}\left[1 - \frac{\pi}{4}\right] = \frac{\pi}{8} - \frac{1}{2} + \frac{\pi}{8} = \frac{\pi}{4} - \frac{1}{2}.$$

3. $\displaystyle\int_0^\pi e^x\cos x\,dx$ applying *type 3* partial integration scenario:

$$\boxed{\begin{array}{l} u = e^x, \qquad\quad du = e^x\,dx \\ dv = \cos x\,dx, \ v = \int \cos x\,dx = \sin x \end{array}};$$

$$= e^x\sin x\Big|_0^\pi - \int_0^\pi e^x\sin x\,dx \quad \boxed{\begin{array}{l} u = e^x, \qquad\quad du = e^x\,dx \\ dv = \sin x\,dx, \ v = \int \sin x\,dx = -\cos x \end{array}};$$

$$= 0 - \left[-e^x\cos x\Big|_0^\pi + \int_0^\pi e^x\cos x\,dx\right] = -e^\pi - 1 - \int_0^\pi e^x\cos x\,dx.$$

Thus, we arrive at the following equation for the unknown integral:

$$\int_0^\pi e^x\cos x\,dx = -e^\pi - 1 - \int_0^\pi e^x\cos x\,dx,$$

solving which, we obtain

$$\int_0^\pi e^x \cos x \, dx = -\frac{1}{2}(e^\pi + 1).$$

4.3. Combining Substitution and Partial Integration

Integration often requires combining the methods of *substitution* and *partial integration*.

Examples 4.8 (Combining Substitution and Partial Integration).

1. $\displaystyle\int x^3 \sin x^2 \, dx$ \hfill *splitting x off x^3;*

$\displaystyle= \int x^2 \sin x^2 \cdot x \, dx$ \hfill *introducing the missing constant 2;*

$\displaystyle= \frac{1}{2}\int x^2 \sin x^2 \cdot 2x \, dx$ \hfill *substituting:* $\boxed{y = x^2, \ dy = 2x\, dx}$;

$\displaystyle= \frac{1}{2}\int y \sin y \, dy$ \hfill applying *type* 1 partial integration scenario:

$$\boxed{\begin{aligned} u &= y, & du &= dy \\ dv &= \sin y \, dy, \ v &= \textstyle\int \sin y \, dy = -\cos y \end{aligned}};$$

$\displaystyle= \frac{1}{2}\left[-y\cos y + \int \cos y \, dy\right] = \frac{1}{2}\left(-y\cos y + \sin y\right) + C$

\hfill *substituting back;*

$\displaystyle= \frac{1}{2}(-x^2 \cos x^2 + \sin x^2) + C.$

2. $\displaystyle\int e^{\sqrt{x}}\, dx$ \hfill *substituting:* $\boxed{y = \sqrt{x}, \ x = y^2, \ dx = 2y\,dy}$;

$\displaystyle= 2\int y e^y \, dy$

applying *type* 1 partial integration scenario: $\boxed{\begin{aligned} u &= y, & du &= dy \\ dv &= e^y \, dy, \ v &= e^y \end{aligned}}$;

$\displaystyle= 2\left[ye^y - \int e^y \, dy\right] = 2\left[ye^y - e^y\right] + C$ \hfill *substituting back;*

$$= 2(\sqrt{x}e^{\sqrt{x}} - e^{\sqrt{x}}) + C = 2(\sqrt{x} - 1)e^{\sqrt{x}} + C.$$

3. $\displaystyle\int \arcsin x\, dx$ applying *type* 2 partial integration scenario:

$$\boxed{\begin{aligned} u &= \arcsin x, \quad du = \frac{1}{\sqrt{1-x^2}}\,dx \\ dv &= dx, \qquad v = x \end{aligned}}\;;$$

$\displaystyle = x\arcsin x - \int \frac{x}{\sqrt{1-x^2}}\,dx$ introducing the *missing constant* -2;

$\displaystyle = x\arcsin x + \frac{1}{2}\int \frac{1}{\sqrt{1-x^2}}(-2x)\,dx$

         substituting: $\boxed{y = 1-x^2,\ dy = -2x\,dx}$;

$\displaystyle = x\arcsin x + \frac{1}{2}\int y^{-1/2}\,dy = x\arcsin x + y^{1/2}$ substituting back;

$\displaystyle = x\arcsin x + \sqrt{1-x^2} + C.$

4.4. Applications

Example 4.9 (Application).
Find the *volume* of the solid obtained by rotating the region bounded by $y = \cos x$ and the x-axis on the interval $[0, \pi/2]$ about the y-axis.

Solution: Since the rotation axis is perpendicular to the axis of definition, by the *shell method*,

$$V = \int_0^{\pi/2} 2\pi x \cos x\, dx = 2\pi \int_0^{\pi/2} x \cos x\, dx$$

         applying *type* 1 partial integration scenario:

$$\boxed{\begin{aligned} u &= x, \qquad\qquad du = dx \\ dv &= \cos x\,dx,\ v = \textstyle\int \cos x\,dx = \sin x \end{aligned}}\;;$$

$\displaystyle = 2\pi \left[x\sin x\Big|_0^{\pi/2} - \int_0^{\pi/2} \sin x\, dx \right]$ by the *Newton-Leibniz Formula*;

$\displaystyle = 2\pi \left[(\pi/2)\sin(\pi/2) + \cos x\Big|_0^{\pi/2} \right] = 2\pi\left[\pi/2 + \cos(\pi/2) - \cos 0\right]$

$$= 2\pi \left[\pi/2 - 1\right] \text{ un.}^3.$$

4.5.   Practice Problems

Evaluate the integrals.

1. $\displaystyle\int x^3 e^{-x}\, dx$

2. $\displaystyle\int x^2 \cos x\, dx$

3. $\displaystyle\int \ln x\, dx$

4. $\displaystyle\int x \operatorname{arcsec} x\, dx$

5. $\displaystyle\int e^{2x} \sin 3x\, dx$

6. $\displaystyle\int \frac{\ln x}{x^2}\, dx$

7. $\displaystyle\int x \sin x \cos x\, dx$

8. $\displaystyle\int \arctan x\, dx$

9. $\displaystyle\int \cos(\ln x)\, dx$

10. $\displaystyle\int (\ln x)^2\, dx$

11. $\displaystyle\int \frac{x}{\sin^2 x}\, dx$

12. $\displaystyle\int x^3 e^{-x^2}\, dx$

13. $\displaystyle\int x(\arctan x)^2\, dx$

14. $\displaystyle\int \frac{\arcsin x}{x^2}\, dx$

15. $\displaystyle\int_0^{\ln 3} x e^x\, dx$

16. $\displaystyle\int_0^{\pi/2} x \cos 2x\, dx$

17. $\displaystyle\int_0^{\pi} e^x \sin x\, dx$

18. $\displaystyle\int_1^{e^2} \sqrt{x}\, \ln x\, dx$

19. $\displaystyle\int_{\pi/6}^{\pi/2} \frac{x \cos x}{\sin^3 x}\, dx$

20. $\displaystyle\int_{2/\sqrt{3}}^{2} x \operatorname{arccsc} x\, dx$

Chapter 5

Trigonometric Integrals

Here, we are to consider certain approaches to finding *trigonometric integrals*, i.e., integrals containing trigonometric functions, e.g.,

$$\int \sin^5 x \sqrt{\cos x}\, dx.$$

5.1. Direct Integration

There are many trigonometric integrals, whose evaluation rests exclusively upon the *integration rules* and relevant trigonometric identities (see Appendix C) and makes no use of *substitution* and *partial integration*. Several such examples and types are considered in Chapter 2 (see Examples 2.6, 2.7, 2.8, and 2.9) and here is one more instance of *direct integration*.

Example 5.1 (Direct Integration).

$$\int \frac{1}{1 - \sin x}\, dx$$

multiplying and dividing by the *conjugate expression* $1 + \sin x$;

$$= \int \frac{1 + \sin x}{(1 - \sin x)(1 + \sin x)}\, dx = \int \frac{1 + \sin x}{1 - \sin^2 x}\, dx$$

using the *trigonometric identity*: $\boxed{1 - \sin^2 x = \cos^2 x}$;

$$= \int \frac{1 + \sin x}{\cos^2 x}\, dx \qquad \text{dividing termwise and rewriting equivalently;}$$

$$= \int \left[\frac{1}{\cos^2 x} + \frac{\sin x}{\cos^2 x} \right] dx = \int \left[\left(\frac{1}{\cos x} \right)^2 + \frac{1}{\cos x} \cdot \frac{\sin x}{\cos x} \right] dx$$

using the *trigonometric identities* again: $\boxed{\dfrac{1}{\cos x} = \sec x, \quad \dfrac{\sin x}{\cos x} = \tan x}$;

$$= \int \left[\sec^2 x + \sec x \tan x \right] dx \qquad\qquad \text{by the } integration \ rules;$$

$$= \int \sec^2 x \, dx + \int \sec x \tan x \, dx = \tan x + \sec x + C.$$

5.2. Using Integration Methods

Not all trigonometric integrals can be evaluated *directly*. There are many relying on the use of *substitution* and *partial integration*. Several such examples and types are considered in Chapter 3 (see Examples 3.1, 3.2, 3.3, 3.5, and 3.6) and Chapter 4 (see Examples 4.3, 4.4, 4.7, 4.8, and 4.9) and more are to be considered here.

5.2.1. *Integration via Reduction Formulas*

As shown in Example 2.7, the trigonometric integral $\int \cos^2 x \, dx$ can be evaluated directly via an appropriate power-reduction identity (see Appendix C).
What should we do if we are to evaluate

$$\int \cos^n x \, dx,$$

where the exponent n is a sufficiently large positive integer?
Such integrals can be evaluated by special *reduction formulas* obtained via *partial integration* or *substitution*.

Theorem 5.1 (Reduction Formulas for Trigonometric Integrals).
For $n = 2, 3, \ldots$,

1. $\displaystyle \int \cos^n x \, dx = \frac{\cos^{n-1} x \sin x}{n} + \frac{n-1}{n} \int \cos^{n-2} x \, dx$

2. $\displaystyle \int \sin^n x \, dx = -\frac{\sin^{n-1} x \cos x}{n} + \frac{n-1}{n} \int \sin^{n-2} x \, dx$

3. $\displaystyle \int \sec^n x \, dx = \frac{\sec^{n-2} x \tan x}{n-1} + \frac{n-2}{n-1} \int \sec^{n-2} x \, dx$

4. $\displaystyle \int \csc^n x \, dx = -\frac{\csc^{n-2} x \cot x}{n-1} + \frac{n-2}{n-1} \int \csc^{n-2} x \, dx$

5. $\displaystyle \int \tan^n x \, dx = \frac{\tan^{n-1} x}{n-1} - \int \tan^{n-2} x \, dx$

6. $\displaystyle\int \cot^n x\, dx = -\frac{\cot^{n-1} x}{n-1} - \int \cot^{n-2} x\, dx$

Proof. Here, we prove the reduction formulas for

$$\int \cos^n x\, dx, \quad \int \sec^n x\, dx, \quad \text{and} \quad \int \tan^n x\, dx.$$

Similar proofs of the remaining ones are left as an *exercise* to the reader.

$\displaystyle\int \cos^n x\, dx$ splitting $\cos x$ off $\cos^n x$;

$\displaystyle= \int \cos^{n-1} x \cos x\, dx$ *integrating by parts:*

$$\boxed{\begin{array}{l} u = \cos^{n-1} x,\ du = (n-1)\cos^{n-2} x(-\sin x)dx \\ dv = \cos x dx,\ v = \int \cos x\, dx = \sin x \end{array}};$$

$\displaystyle= \cos^{n-1} x \sin x + (n-1)\int \cos^{n-2} x \sin^2 x\, dx$

by the *trigonometric identity*: $\boxed{\sin^2 x = 1 - \cos^2 x}$;

$\displaystyle= \cos^{n-1} x \sin x + (n-1)\int \cos^{n-2} x(1 - \cos^2 x)\, dx$ multiplying;

$\displaystyle= \cos^{n-1} x \sin x + (n-1)\int \left[\cos^{n-2} x - \cos^n x\right] dx$

by the *integration rules*;

$\displaystyle= \cos^{n-1} x \sin x + (n-1)\int \cos^{n-2} x\, dx - (n-1)\int \cos^n x\, dx.$

Thus, we arrive at the following equation for the initial integral:

$$\int \cos^n x\, dx = \cos^{n-1} x \sin x + (n-1)\int \cos^{n-2} x\, dx - (n-1)\int \cos^n x\, dx,$$

solving which we obtain the reduction formula

$$\int \cos^n x\, dx = \frac{\cos^{n-1} x \sin x}{n} + \frac{n-1}{n}\int \cos^{n-2} x\, dx.$$

Similarly,

$\displaystyle\int \sec^n x\, dx$ splitting $\sec^2 x$ off $\sec^n x$;

$$= \int \sec^{n-2} x \sec^2 x \, dx \qquad\qquad \textit{integrating by parts:}$$

$$\boxed{\begin{aligned} u &= \sec^{n-2} x, \quad du = (n-2)\sec^{n-3} x \sec x \tan x \, dx \\ dv &= \sec^2 x \, dx, \ v = \int \sec^2 x \, dx = \tan x \end{aligned}};$$

$$= \sec^{n-2} x \tan x - (n-2) \int \sec^{n-2} x \tan^2 x \, dx$$

$$\textit{by the trigonometric identity:} \quad \boxed{\tan^2 x = \sec^2 x - 1};$$

$$= \sec^{n-2} x \tan x - (n-2) \int \sec^{n-2} x (\sec^2 x - 1) \, dx$$

$$\textit{by the integration rules};$$

$$= \sec^{n-2} x \tan x - (n-2) \int \sec^n x \, dx + (n-2) \int \sec^{n-2} x \, dx.$$

Thus, we arrive at the following equation for the initial integral:

$$\int \sec^n x \, dx = \sec^{n-2} x \tan x - (n-2) \int \sec^n x \, dx + (n-2) \int \sec^{n-2} x \, dx,$$

solving which we obtain the reduction formula

$$\int \sec^n x \, dx = \frac{\sec^{n-2} x \tan x}{n-1} + \frac{n-2}{n-1} \int \sec^{n-2} x \, dx.$$

The next reduction formula is obtained via *substitution*:

$$\int \tan^n x \, dx \qquad\qquad \textit{splitting } \tan^2 x \textit{ off } \tan^n x;$$

$$= \int \tan^{n-2} x \tan^2 x \, dx$$

$$\textit{by the trigonometric identity:} \quad \boxed{\tan^2 x = \sec^2 x - 1};$$

$$= \int \tan^{n-2} x (\sec^2 x - 1) \, dx \qquad \textit{by the integration rules};$$

$$= \int \tan^{n-2} x \sec^2 x \, dx - \int \tan^{n-2} x \, dx$$

$$\textit{substituting for the first integral:} \quad \boxed{u = \tan x, \quad du = \sec^2 x \, dx};$$

$$= \int u^{n-2} \, du - \int \tan^{n-2} x \, dx = \frac{u^{n-1}}{n-1} - \int \tan^{n-2} x \, dx$$

$$\textit{substituting back};$$

$$= \frac{\tan^{n-1} x}{n-1} - \int \tan^{n-2} x \, dx.$$

$\square$

Remarks 5.1 (Reduction Formulas).

- If the exponent $n = 2k$ is a positive *even* integer, applying the corresponding reduction formula precisely k times, we arrive at

$$\int \cos^0 x \, dx = \int \sin^0 x \, dx = \int \sec^0 x \, dx = \int \csc^0 x \, dx$$

$$= \int \tan^0 x \, dx = \int \cot^0 x \, dx = \int 1 \, dx = x + C.$$

- If the exponent $n = 2k + 1$ is a positive *odd* integer, applying the corresponding reduction formula precisely k times, we arrive at

$$\int \cos x \, dx = \sin x + C, \quad \int \sin x \, dx = -\cos x + C,$$

$$\int \sec x \, dx = \ln|\sec x + \tan x| + C, \quad \int \csc x \, dx = \ln|\csc x - \cot x| + C,$$

$$\int \tan x \, dx = \ln|\sec x| + C, \quad \text{or} \quad \int \cot x \, dx = \ln|\sin x| + C.$$

- Observe that, whenever the reduction formulas for

$$\int \cos^n x \, dx, \quad \int \sin^n x \, dx, \quad \int \sec^n x \, dx, \quad \text{or} \quad \int \csc^n x \, dx$$

are used, *partial integration* is implicitly executed and, whenever the reduction formulas for

$$\int \tan^n x \, dx, \quad \text{or} \quad \int \cot^n x \, dx$$

are used, *substitution* is implicitly executed.

Examples 5.2 (Using Reduction Formulas).

1. $\displaystyle\int \cos^5 x \, dx$ applying the appropriate reduction formula *twice*;

$$= \frac{\cos^4 x \sin x}{5} + \frac{4}{5} \int \cos^3 x \, dx = \frac{\cos^4 x \sin x}{5} + \frac{4}{5} \left[\frac{\cos^2 x \sin x}{3} \right.$$

$$+ \frac{2}{3} \int \cos x \, dx \Bigg] = \frac{1}{5} \cos^4 x \sin x + \frac{4}{15} \cos^2 x \sin x + \frac{8}{15} \sin x + C.$$

2. $\int \tan^6 x \, dx$ applying the appropriate reduction formula *threefold*;

$$= \frac{\tan^5 x}{5} - \int \tan^4 x \, dx = \frac{\tan^5 x}{5} - \left[\frac{\tan^3 x}{3} - \int \tan^2 x \, dx \right]$$

$$= \frac{\tan^5 x}{5} - \frac{\tan^3 x}{3} + \tan x - \int 1 \, dx = \frac{\tan^5 x}{5} - \frac{\tan^3 x}{3} + \tan x - x + C.$$

5.2.2.   Integrals of the Form $\int \sin^m x \cos^n x \, dx$

When evaluating a trigonometric integral of the form

$$\int \sin^m x \cos^n x \, dx,$$

where m and n are real exponents, we may encounter three special cases allowing clear-cut integration strategies. They, however, do not encompass all the possibilities and do not apply to, say,

$$\int \sin^{\sqrt{2}} x \cos^\pi x \, dx.$$

The three special cases are as follows.

- **Case 1:** m is an *odd positive integer*, i.e., $m = 2k + 1$ with some $k = 0, 1, \ldots$ and n is an *arbitrary real number*. In this case, we use the following integration strategy:

    Step 1: split $\sin x$ off $\sin^{2k+1} x$;
    Step 2: express the remaining *even* power $\sin^{2k} x$ in terms of $\cos x$ via the identity $\boxed{\sin^2 x = 1 - \cos^2 x}$:
    $$\sin^{2k} x = [\sin^2 x]^k = [1 - \cos^2 x]^k;$$
    Step 3: integrate via the *substitution* $u = \cos x$.

Example 5.3.

$$\int \sin^5 x \sqrt{\cos x} \, dx \qquad\qquad \text{switching to the } power\ form;$$

$$= \int \sin^5 x \cos^{1/2} x \, dx \quad (m = 5,\ n = 1/2) \quad splitting \sin x \text{ off } \sin^5 x;$$

$$= \int \sin^4 x \cos^{1/2} x \sin x \, dx$$

expressing $\sin^4 x$ in terms of $\cos x$ via $\boxed{\sin^2 x = 1 - \cos^2 x}$;

$$= \int (1 - \cos^2 x)^2 \cos^{1/2} x \sin x \, dx$$

introducing the *missing constant* -1;

$$= -\int (1 - \cos^2 x)^2 \cos^{1/2} x (-\sin x) \, dx$$

substituting: $\boxed{u = \cos x, \ du = -\sin x \, dx}$;

$$= -\int (1 - u^2)^2 u^{1/2} \, du$$

integrating directly relative to the new variable;

$$= -\int (1 - 2u^2 + u^4) u^{1/2} \, du = -\int (u^{1/2} - 2u^{5/2} + u^{9/2}) \, du$$

$$= -\left[\int u^{1/2} \, du - 2 \int u^{5/2} \, du + \int u^{9/2} \, du \right]$$

$$= -\left[\frac{2}{3} u^{3/2} - \frac{4}{7} u^{7/2} + \frac{2}{11} u^{11/2} \right] + C$$

substituting back and distributing;

$$= -\frac{2}{3} \cos^{3/2} x + \frac{4}{7} \cos^{7/2} x - \frac{2}{11} \cos^{11/2} x + C.$$

- **Case 2:** n is an *odd positive integer*, i.e., $n = 2k + 1$ with some $k = 0, 1, \ldots$ and m is an *arbitrary real number*. In this case, we use the following integration strategy:

  Step 1: split $\cos x$ off $\cos^{2k+1} x$;
  Step 2: express the remaining *even* power $\cos^{2k} x$ in terms of $\sin x$ via the identity $\boxed{\cos^2 x = 1 - \sin^2 x}$:

  $$\cos^{2k} x = [\cos^2 x]^k = [1 - \sin^2 x]^k;$$

  Step 3: integrate via the *substitution* $u = \sin x$.

Example 5.4.

$$\int \sin^4 x \cos^3 x \, dx \qquad (m = 4, \ n = 3) \quad \text{splitting } \cos x \text{ off } \cos^3 x;$$

$$= \int \sin^4 x \cos^2 x \cos x \, dx$$

expressing $\cos^2 x$ in terms of $\sin x$ via $\boxed{\cos^2 x = 1 - \sin^2 x}$;

$$= \int \sin^4 x (1 - \sin^2 x) \cos x \, dx$$

substituting: $\boxed{u = \sin x, \ du = \cos x \, dx}$

$$= \int u^4 (1 - u^2) \, du$$

multiplying and integrating relative to the new variable;

$$= \int (u^4 - u^6) \, du = \int u^4 \, du - \int u^6 \, du = \frac{u^5}{5} - \frac{u^7}{7} + C$$

substituting back;

$$= \frac{\sin^5 x}{5} - \frac{\sin^7 x}{7} + C.$$

In Examples 5.2, the integral

$$\int \cos^5 x \, dx$$

is evaluated via the appropriate *reduction formula*. Now, let us apply the described new strategy to it.

Example 5.5 (Alternate Approach).

$$\int \cos^5 x \, dx \qquad\qquad (m = 0, \ n = 5) \quad \textit{splitting } \cos x \textit{ off } \cos^5 x;$$

$$= \int \cos^4 x \cos x \, dx$$

expressing $\cos^4 x$ in terms of $\sin x$ via $\boxed{\cos^2 x = 1 - \sin^2 x}$;

$$= \int (1 - \sin^2 x)^2 \cos x \, dx$$

substituting: $\boxed{u = \sin x, \ du = \cos x \, dx}$;

$$= \int (1 - u^2)^2 \, du$$

expanding and integrating relative to the new variable;

$$= \int (1 - 2u^2 + u^4) \, du = \int 1 \, du - 2 \int u^2 \, du + \int u^4 \, du$$

$$= u - 2\frac{u^3}{3} + \frac{u^5}{5} + C \qquad\qquad \text{substituting back;}$$

$$= \sin x - \frac{2}{3}\sin^3 x + \frac{1}{5}\sin^5 x + C.$$

Remark 5.2. Although the answers afforded by the two different approaches look very different (cf. Examples 5.2), they actually differ by a constant only.

- **Case 3:** both exponents m and n are *even nonnegative integers*, i.e., $m = 2k$ and $n = 2l$ with $k, l = 0, 1, \ldots$. In this case, we use the following integration strategy:

  Step 1: use the *power reduction identities*
  $$\cos^2 x = \frac{1 + \cos 2x}{2}, \quad \sin^2 x = \frac{1 - \cos 2x}{2}$$
  to transform the integrand into a polynomial in $\cos 2x$;

  Step 2: substitute $u = 2x$ and apply the appropriate *reduction formula* to the powers of $\cos u$ greater than 1 or use the preceding strategies, i.e., repeatedly apply the *power reduction identity* to the *even* powers of $\cos 2x$ and *substitution* to the odd ones.

Example 5.6 (Via Power Reduction and Reduction Formula).

$$\int \sin^4 x \cos^2 x \, dx \qquad\qquad (m = 4, \ n = 2)$$

by the *power reduction identities*:

$$\boxed{\cos^2 x = \frac{1 + \cos 2x}{2}, \quad \sin^2 x = \frac{1 - \cos 2x}{2}};$$

$$= \int \left(\frac{1 - \cos 2x}{2}\right)^2 \frac{1 + \cos 2x}{2} \, dx = \int \frac{1}{8}(1 - \cos 2x)^2 (1 + \cos 2x) \, dx$$

multiplying and simplifying;

$$= \int \frac{1}{8}(1 - 2\cos 2x + \cos^2 2x)(1 + \cos 2x) \, dx$$

$$= \int \frac{1}{8}(1 - \cos 2x - \cos^2 2x + \cos^3 2x) \, dx$$

by the *integration rules*;

$$= \frac{1}{8}\left[\int 1 \, dx - \int \cos 2x \, dx - \int \cos^2 2x \, dx + \int \cos^3 2x \, dx\right]$$

$$= \frac{1}{8}x - \frac{1}{16}\sin 2x - \frac{1}{8}\int \cos^2 2x\, dx + \frac{1}{8}\int \cos^3 2x\, dx$$

$$\text{substituting:}\quad \boxed{u = 2x,\ du = 2dx,\ dx = \frac{1}{2}\,du};$$

$$= \frac{1}{8}x - \frac{1}{16}\sin 2x - \frac{1}{16}\int \cos^2 u\, du + \frac{1}{16}\int \cos^3 u\, du$$

applying the appropriate *reduction formula* to $\cos^2 u$ and $\cos^3 u$;

$$= \frac{1}{8}x - \frac{1}{16}\sin 2x - \frac{1}{16}\left[\frac{\cos u \sin u}{2} + \frac{1}{2}\int 1\, du\right]$$

$$+ \frac{1}{16}\left[\frac{\cos^2 u \sin u}{3} + \frac{2}{3}\int \cos u\, du\right]$$

$$= \frac{1}{8}x - \frac{1}{16}\sin 2x - \frac{1}{32}\cos u \sin u - \frac{1}{32}u + \frac{1}{48}\cos^2 u \sin u + \frac{1}{24}\sin u + C$$

$$\text{substituting back;}$$

$$= \frac{1}{8}x - \frac{1}{16}\sin 2x - \frac{1}{32}\cos 2x \sin 2x - \frac{1}{16}x + \frac{1}{48}\cos^2 2x \sin 2x$$

$$+ \frac{1}{24}\sin 2x + C \qquad\qquad\qquad\qquad\qquad\qquad \text{simplifying;}$$

$$= \frac{1}{16}x - \frac{1}{48}\sin 2x - \frac{1}{32}\cos 2x \sin 2x + \frac{1}{48}\cos^2 2x \sin 2x + C.$$

Example 5.7 (Via Power Reduction and Substitution).

$$\int \sin^4 x \cos^2 x\, dx \qquad\qquad \text{by the *power reduction identities*:}$$

$$\boxed{\cos^2 x = \frac{1 + \cos 2x}{2}, \quad \sin^2 x = \frac{1 - \cos 2x}{2}};$$

$$= \int \left(\frac{1 - \cos 2x}{2}\right)^2 \frac{1 + \cos 2x}{2}\, dx = \int \frac{1}{8}(1 - \cos 2x)^2(1 + \cos 2x)\, dx$$

$$\text{multiplying and simplifying;}$$

$$= \frac{1}{8}\int (1 - 2\cos 2x + \cos^2 2x)(1 + \cos 2x)\, dx$$

$$= \frac{1}{8}\int (1 - \cos 2x - \cos^2 2x + \cos^3 2x)\, dx$$

$$\text{by the *integration rules*;}$$

$$= \frac{1}{8}\left[\int 1\, dx - \int \cos 2x\, dx - \int \cos^2 2x\, dx + \int \cos^3 2x\, dx\right]$$

$$= \frac{1}{8}x - \frac{1}{16}\sin 2x - \frac{1}{8}\int \cos^2 2x\,dx + \frac{1}{8}\int \cos^3 2x\,dx$$

applying the *power reduction identity*
and *splitting* $\cos 2x$ off $\cos^3 2x$;

$$= \frac{1}{8}x - \frac{1}{16}\sin 2x - \frac{1}{8}\int \frac{1+\cos 4x}{2}\,dx + \frac{1}{8}\int \cos^2 2x \cos 2x\,dx$$

expressing $\cos^2 2x$ in terms of $\sin 2x$;

$$= \frac{1}{8}x - \frac{1}{16}\sin 2x - \frac{1}{8}\int \frac{1+\cos 4x}{2}\,dx + \frac{1}{8}\int [1-\sin^2 2x]\cos 2x\,dx$$

by the *integration rules* and introducing the *missing constant* 2;

$$= \frac{1}{8}x - \frac{1}{16}\sin 2x - \frac{1}{16}\left[\int 1\,dx + \int \cos 4x\,dx\right]$$
$$+ \frac{1}{8}\cdot\frac{1}{2}\int [1-\sin^2 2x](2\cos 2x)\,dx$$

substituting: $\boxed{u = \sin 2x,\ du = 2\cos 2x\,dx}$;

$$= \frac{1}{8}x - \frac{1}{16}\sin 2x - \frac{1}{16}x - \frac{1}{64}\sin 4x + \frac{1}{16}\int [1-u^2]\,du$$

integrating relative to the new variable;

$$= \frac{1}{16}x - \frac{1}{16}\sin 2x - \frac{1}{64}\sin 4x + \frac{1}{16}\left[u - \frac{u^3}{3}\right] + C$$

substituting back;

$$= \frac{1}{16}x - \frac{1}{16}\sin 2x - \frac{1}{64}\sin 4x + \frac{1}{16}\left[\sin 2x - \frac{1}{3}\sin^3 2x\right] + C$$

simplifying;

$$= \frac{1}{16}x - \frac{1}{64}\sin 4x - \frac{1}{48}\sin^3 2x + C.$$

Remark 5.3. There may also exist an approach based exclusively on *trigonometric identities*.

Example 5.8 (Via Trigonometric Identities Only).

$$\int \sin^4 x \cos^2 x\,dx$$

splitting $\sin^2 x$ off $\sin^4 x$ and combining with $\cos^2 x$;

$$= \int \sin^2 x [\sin x \cos x]^2 \, dx$$

by the *double-angle identity*: $\boxed{\sin x \cos x = \dfrac{1}{2} \sin 2x}$;

$$= \frac{1}{4} \int \sin^2 x \sin^2 2x \, dx$$

by the *power reduction identity*: $\boxed{\sin^2 \theta = \dfrac{1 - \cos 2\theta}{2}}$;

$$= \frac{1}{16} \int (1 - \cos 2x)(1 - \cos 4x) \, dx \qquad \text{multiplying;}$$

$$= \frac{1}{16} \int (1 - \cos 2x - \cos 4x + \cos 2x \cos 4x) \, dx \quad \text{by the } \textit{integration rules;}$$

$$= \frac{1}{16} \int 1 \, dx - \frac{1}{16} \int \cos 2x \, dx - \frac{1}{16} \int \cos 4x \, dx + \frac{1}{16} \int \cos 4x \cos 2x \, dx$$

by the *product-to-sum identity*:

$$\boxed{\cos \alpha \cos \beta = \frac{1}{2} [\cos(\alpha - \beta) + \cos(\alpha + \beta)]};$$

$$= \frac{1}{16} x - \frac{1}{32} \sin 2x - \frac{1}{64} \sin 4x + \frac{1}{32} \int (\cos 2x + \cos 6x) \, dx$$

by the *integration rules;*

$$= \frac{1}{16} x - \frac{1}{32} \sin 2x - \frac{1}{64} \sin 4x + \frac{1}{32} \left[\int \cos 2x \, dx + \int \cos 6x \, dx \right]$$

$$= \frac{1}{16} x - \frac{1}{32} \sin 2x - \frac{1}{64} \sin 4x + \frac{1}{64} \sin 2x + \frac{1}{192} \sin 6x + C$$

$$= \frac{1}{16} x - \frac{1}{64} \sin 2x - \frac{1}{64} \sin 4x + \frac{1}{192} \sin 6x + C.$$

Remark 5.4. Although the answers afforded by the three different approaches look very different, they essentially differ from each other by a constant only.

5.2.3. *Integrals of the Form* $\int \tan^m x \sec^n x \, dx$

When evaluating a trigonometric integral of the form

$$\int \tan^m x \sec^n x \, dx,$$

where m and n are real exponents, similarly to

$$\int \sin^m x \cos^n x \, dx,$$

we may encounter three special cases allowing clear-cut integration strategies. They do not encompass all the possibilities and do not apply to, say,

$$\int \tan^\pi x \sec^{\sqrt{2}} x \, dx.$$

The three special cases are as follows.

- **Case 1:** n is an *even positive integer*, i.e., $n = 2k+2$ with $k = 0, 1, \ldots$ and m is an *arbitrary real number*. In this case, we use the following integration strategy:

  Step 1: split $\sec^2 x$ off $\sec^{2k+2} x$;

  Step 2: express the remaining *even* power $\sec^{2k} x$ in terms of $\tan x$ via the identity $\boxed{\sec^2 x = \tan^2 x + 1}$:
  $$\sec^{2k} x = [\sec^2 x]^k = [\tan^2 x + 1]^k;$$

  Step 3: integrate via the *substitution* $u = \tan x$.

Example 5.9.

$$\int \tan^3 x \sec^4 x \, dx \qquad (m = 3, \ n = 4) \quad \textit{splitting } \sec^2 x \textit{ off } \sec^4 x;$$

$$= \int \tan^3 x \sec^2 x \sec^2 x \, dx$$

$$\textit{expressing } \sec^2 x \textit{ in terms of } \tan x \textit{ via } \boxed{\sec^2 x = \tan^2 x + 1};$$

$$= \int \tan^3 x (\tan^2 x + 1) \sec^2 x \, dx$$

$$\textit{substituting: } \boxed{u = \tan x, \ du = \sec^2 x \, dx};$$

$$= \int u^3 (u^2 + 1) \, du$$

$$\textit{multiplying and integrating relative to the new variable;}$$

$$= \int (u^5 + u^3) \, du = \int u^5 \, du + \int u^3 \, du = \frac{u^6}{6} + \frac{u^4}{4} + C$$

$$\textit{substituting back;}$$

$$= \frac{\tan^6 x}{6} + \frac{\tan^4 x}{4} + C.$$

- **Case 2:** m is an *odd positive integer*, i.e., $m = 2k + 1$ with $k = 0, 1, \ldots$ and n is an *arbitrary real number*. In this case, we use the following integration strategy:

  Step 1: split $\sec x \tan x$ off $\tan^{2k+1} x \sec^n x$;
  Step 2: express the remaining *even* power $\tan^{2k} x$ in terms of $\sec x$ via the identity $\boxed{\tan^2 x = \sec^2 x - 1}$:

  $$\tan^{2k} x = [\tan^2 x]^k = [\sec^2 x - 1]^k;$$

  Step 3: integrate via the *substitution* $u = \sec x$.

Example 5.10.

$$\int \tan^3 x \sec^5 x \, dx \, (m = 3, \ n = 5) \quad \textit{splitting } \sec x \tan x \textit{ off } \tan^3 x \sec^5 x;$$

$$= \int \tan^2 x \sec^4 (\sec x \tan x) \, dx$$

$$\textit{expressing } \tan^2 x \textit{ in terms of } \sec x \textit{ via } \boxed{\tan^2 x = \sec^2 x - 1};$$

$$= \int (\sec^2 x - 1) \sec^4 x (\sec x \tan x) \, dx$$

$$\textit{substituting: } \boxed{u = \sec x, \ du = \sec x \tan x \, dx};$$

$$= \int (u^2 - 1) u^4 \, du$$

$$\textit{multiplying and integrating relative to the new variable;}$$

$$= \int u^6 \, du - \int u^4 \, du = \frac{u^7}{7} - \frac{u^5}{5} + C \qquad \textit{substituting back;}$$

$$= \frac{\sec^7 x}{7} - \frac{\sec^5 x}{5} + C.$$

- **Case 3:** m is an *even nonnegative integer*, i.e., $m = 2k$ with $k = 0, 1, \ldots$, and n is an *odd positive integer*, i.e., $n = 2l + 1$ with $l = 0, 1, \ldots$. In this case, we use the following integration strategy:

  Step 1: express the *even* power $\tan^{2k} x$ in terms of $\sec x$ via the identity $\boxed{\tan^2 x = \sec^2 x - 1}$:

  $$\tan^{2k} x = [\tan^2 x]^k = [\sec^2 x - 1]^k,$$

  to transform the integrand into a polynomial in $\sec x$;

Step 2: apply the appropriate *reduction formula* to the powers of $\sec x$ greater than 1.

Example 5.11.

$$\int \tan^2 x \sec x \, dx \qquad\qquad (m = 2, \ n = 1)$$

expressing $\tan^2 x$ in terms of $\sec x$ via $\boxed{\tan^2 x = \sec^2 x - 1}$;

$$= \int (\sec^2 x - 1) \sec x \, dx \qquad\qquad \text{multiplying;}$$

$$= \int (\sec^3 x - \sec x) \, dx = \int \sec^3 x \, dx - \int \sec x \, dx$$

applying the appropriate *reduction formula* to the first integral;

$$= \frac{\sec x \tan x}{2} + \frac{1}{2} \int \sec x \, dx - \int \sec x \, dx$$

$$= \frac{\sec x \tan x}{2} - \frac{1}{2} \ln |\sec x + \tan x| + C.$$

Remark 5.5. Similar strategies, in the corresponding cases, apply to integrals of the form

$$\int \cot^m x \csc^n x \, dx.$$

5.3. Applications

Examples 5.12 (Applications).

1. Find the *volume* of the solid obtained by rotating the region bounded by $y = \sin^2 x$ and the x-axis on the interval $[0, \pi/2]$ about the x-axis.

   Solution: Applying the *disk method* (see, e.g., [1, 6]), we have:

$$V = \int_0^{\pi/2} \pi \left[\sin^2 x\right]^2 dx$$

by the *power-reduction identity*: $\boxed{\sin^2 x = \dfrac{1 - \cos 2x}{2}}$;

$$= \int_0^{\pi/2} \pi \left[\frac{1 - \cos 2x}{2}\right]^2 dx \qquad\qquad \text{squaring;}$$

$$= \int_0^{\pi/2} \frac{\pi}{4} \left[1 - 2\cos 2x + \cos^2 2x\right] dx$$

$$\text{by the } power\text{-}reduction \text{ identity:} \quad \boxed{\cos^2 2x = \frac{1 + \cos 4x}{2}};$$

$$= \frac{\pi}{4} \int_0^{\pi/2} \left[1 - 2\cos 2x + \frac{1 + \cos 4x}{2} \right] dx \qquad \text{by the } integration \text{ rules;}$$

$$= \frac{\pi}{4} \left[\frac{3}{2} \int_0^{\pi/2} 1 \, dx - 2 \int_0^{\pi/2} \cos 2x + \frac{1}{2} \int_0^{\pi/2} \cos 4x \, dx \right]$$

$$\text{by the } Newton\text{-}Leibniz \text{ Formula;}$$

$$= \frac{\pi}{4} \left[\frac{3}{2} x \Big|_0^{\pi/2} - \sin 2x \Big|_0^{\pi/2} + \frac{1}{2} \cdot \frac{1}{4} \sin 4x \Big|_0^{\pi/2} \right]$$

$$= \frac{\pi}{4} \left[\frac{3}{2} \left(\frac{\pi}{2} - 0 \right) - (\sin \pi - \sin 0) + \frac{1}{8} (\sin 2\pi - \sin 0) \right] = \frac{3\pi^2}{16} \text{ un.}^3.$$

2. Find the *length* of the curve $y = \ln(\sin x)$, $\pi/4 \le x \le \pi/2$.

   Solution: By the *arc length formula*,

$$L = \int_{\pi/4}^{\pi/2} \sqrt{1 + [y'(x)]^2} \, dx \qquad \text{since} \quad y' = \frac{\sin' x}{\sin x} = \frac{\cos x}{\sin x} = \cot x;$$

$$= \int_{\pi/4}^{\pi/2} \sqrt{1 + \cot^2 x} \, dx$$

$$\text{by the trigonometric identity:} \quad \boxed{1 + \cot^2 x = \csc^2 x};$$

$$= \int_{\pi/4}^{\pi/2} \sqrt{\csc^2 x} \, dx \qquad\qquad\qquad \text{considering that}$$

$$\sqrt{\csc^2 x} = |\csc x| = \csc x \text{ since } \csc x > 0 \text{ on } [\pi/4, \pi/2];$$

$$= \int_{\pi/4}^{\pi/2} \csc x \, dx \qquad\qquad \text{by the } Newton\text{-}Leibniz \text{ Formula;}$$

$$= \ln|\csc x - \cot x| \Big|_{\pi/4}^{\pi/2} = \ln|\csc(\pi/2) - \cot(\pi/2)| - \ln|\csc(\pi/4) - \cot(\pi/4)|$$

$$= \ln|1 - 0| - \ln|\sqrt{2} - 1| = -\ln(\sqrt{2} - 1) = -\ln \frac{1}{\sqrt{2} + 1} = \ln(\sqrt{2} + 1) \text{ un.}$$

5.4. Practice Problems

Evaluate the integrals.

1. $\displaystyle\int \sin 5x \cos 2x \, dx$

2. $\displaystyle\int \cos^2 3x \, dx$

3. $\displaystyle\int (1 + 2\sin x)^2 \, dx$

4. $\displaystyle\int (1 - \cos 2x)^2 \, dx$

5. $\displaystyle\int \sin^2 x \cos^2 x \, dx$

6. $\displaystyle\int \sin^2 x \cos^4 x \, dx$

7. $\displaystyle\int \sin^3 x \cos^2 x \, dx$

8. $\displaystyle\int \sin^2 x \cos^3 x \, dx$

9. $\displaystyle\int \sin^7 x \, dx$

10. $\displaystyle\int \cos^4 x \, dx$

11. $\displaystyle\int \frac{\sin^3 x}{\cos^2 x} \, dx$

12. $\displaystyle\int \frac{\cos^3 x}{\sqrt[3]{\sin^2 x}} \, dx$

13. $\displaystyle\int \frac{\sec^4 x}{\tan^3 x} \, dx$

14. $\displaystyle\int \cot^3 x \csc^5 x \, dx$

15. $\displaystyle\int \tan^5 x \, dx$

16. $\displaystyle\int \frac{\sec^3 (\ln x)}{x} \, dx$

17. $\displaystyle\int_{\pi/6}^{\pi/3} \cot^4 x \, dx$

18. $\displaystyle\int_{0}^{\sqrt{\pi/2}} x \cos^3 (x^2) \, dx$

19. $\displaystyle\int_{0}^{\pi/2} \sqrt{1 + \cos 2x} \, dx$

20. $\displaystyle\int_{0}^{\pi} \sin^{2n} x \, dx$ for $n = 1, 2, \ldots$

Chapter 6

Trigonometric Substitutions

6.1. Reverse Substitutions

Frequently, an integral

$$\int f(x)\,dx \quad \text{or} \quad \int_a^b f(x)\,dx$$

can be simplified if we *substitute* a suitably chosen *continuously differentiable* function $g(t)$ for x:

$$\int f(x)\,dx \qquad\qquad \text{with} \quad \boxed{x = g(t),\ dx = g'(t)\,dt};$$

$$= \int f(g(t))g'(t)\,dt,$$

in which case, we use the *Substitution Rule for Indefinite/Definite Integral* (Theorem 3.1/Theorem 3.2) backwards. Thus, we call such a substitution *reverse*.

For the definite integral

$$\int_a^b f(x)\,dx,$$

we also change the integration limits:

$$\int_a^b f(x)\,dx \qquad\qquad \text{with} \quad \boxed{\begin{array}{c|c} x & t \\ \hline a & c \\ b & d \end{array}}\ x = g(t),\ dx = g'(t)\,dt;$$

$$= \int_c^d f(g(t))g'(t)\,dt,$$

Requirements to Reverse Substitution

A *suitable choice* of a reverse substitution $x = g(t)$ is based on the following natural criteria:

1. The substitution should *simplify* the initial integral, i.e., transform it into an integral, which can be evaluated via known integration strategies.
2. The function $g(t)$ must be *continuously differentiable* on some interval.
3. The function $g(t)$ must also be *one-to-one* on its interval of definition so that there exist an *inverse* $g^{-1}(x)$ for us to be able to *substitute back*:

$$t = g^{-1}(x)$$

for indefinite integral or have a *unique choice* for the new integration limits:

x	t
a	$c = g^{-1}(a)$
b	$d = g^{-1}(b)$

Trigonometric Substitutions

Trigonometric substitutions, we are about to consider, are *reverse* by nature and used to transform integrals containing the terms

$$a^2 - x^2, \ x^2 + a^2, \text{ or } x^2 - a^2,$$

where $a > 0$, into *trigonometric integrals*.

6.2. Integrals Containing $a^2 - x^2$

For an integral containing the term $a^2 - x^2$, where $a > 0$, especially in the *radical* form $\sqrt{a^2 - x^2}$, we use the *sine substitution*:

$$x = a\sin\theta, \ -\pi/2 \le \theta \le \pi/2.$$

Remark 6.1. The *range* for the new variable θ is purposely chosen to be $[-\pi/2, \pi/2]$, since

- on $[-\pi/2, \pi/2]$, the function $a\sin\theta$ is *one-to-one*, and thus, we can substitute back:

$$\theta = \arcsin\frac{x}{a},$$

and has the *largest range* of values: the *entire* interval $[-a, a]$;

- also, on $[-\pi/2, \pi/2]$, $\cos\theta \geq 0$, which is instrumental for the *elimination* of the radical $\sqrt{a^2 - x^2}$ (see below).

By the *Pythagorean trigonometric identity*

$$1 - \sin^2\theta = \cos^2\theta$$

(see Appendix C), we have:

$$a^2 - x^2 = a^2 - a^2\sin^2\theta = a^2(1 - \sin^2\theta) = a^2\cos^2\theta.$$

Hence, the substitution eliminates the *radical* $\sqrt{a^2 - x^2}$:

$$\sqrt{a^2 - x^2} = \sqrt{a^2\cos^2\theta} = |a||\cos\theta|$$

considering that $a > 0$ and $\cos\theta \geq 0$ for $-\pi/2 \leq \theta \leq \pi/2$;

$$= a\cos\theta.$$

The substitution transforms the initial integral into a *trigonometric integral* relative to the new variable θ.

Example 6.1 (Integrals Containing $a^2 - x^2$).

$$\int_1^{\sqrt{2}} \frac{\sqrt{2 - x^2}}{x^2}\,dx = \int_1^{\sqrt{2}} \frac{\sqrt{\sqrt{2}^2 - x^2}}{x^2}\,dx$$

substituting and *changing* the *integration limits*:

$x = \sqrt{2}\sin\theta$		
$dx = \sqrt{2}\cos\theta\,d\theta$	x	θ
$\sqrt{2 - x^2} = \sqrt{2}\cos\theta$	1	$\pi/4$
$\theta = \arcsin\dfrac{x}{\sqrt{2}}$	$\sqrt{2}$	$\pi/2$

$$= \int_{\pi/4}^{\pi/2} \frac{\sqrt{2}\cos\theta}{2\sin^2\theta}\sqrt{2}\cos\theta\,d\theta = \int_{\pi/4}^{\pi/2} \frac{\cos^2\theta}{\sin^2\theta}\,d\theta = \int_{\pi/4}^{\pi/2}\left[\frac{\cos\theta}{\sin\theta}\right]^2 d\theta$$

by the *trigonometric identity*: $\boxed{\dfrac{\cos\theta}{\sin\theta} = \cot\theta}$;

$$= \int \cot^2\theta\,d\theta \qquad \text{by the } \textit{trigonometric identity}: \boxed{\cot^2\theta = \csc^2\theta - 1};$$

$$= \int_{\pi/4}^{\pi/2} (\csc^2\theta - 1)\,d\theta \qquad\qquad \text{by the } \textit{integration rules};$$

$$= \int_{\pi/4}^{\pi/2} \csc^2\theta\,d\theta - \int_{\pi/4}^{\pi/2} 1\,d\theta \qquad\qquad \text{by the } \textit{Newton-Leibniz Formula};$$

$$= -\cot\theta\Big|_{\pi/4}^{\pi/2} - \theta\Big|_{\pi/4}^{\pi/2} = -[\cot(\pi/2) - \cot(\pi/4)] - [\pi/2 - \pi/4]$$

$$= -[0-1] - \pi/4 = 1 - \pi/4.$$

What If the Integral Is Indefinite?

If the initial integral is *indefinite*, after having *substituted* and *evaluated* the integral relative to the new variable θ, we obtain an answer in terms of θ and/or some *trigonometric functions* of θ. Thus, when *substituting back*:

$$\theta = \arcsin\frac{x}{a},$$

we are to encounter expressions containing $\arcsin\frac{x}{a}$ inside trigonometric functions, which should be equivalently transformed into algebraic expressions in x not containing the trigonometric functions of $\arcsin\frac{x}{a}$.

This can be done based on *cancellation* and *trigonometric identities* (taking into account that $-\frac{\pi}{2} \le \arcsin\frac{x}{a} \le \frac{\pi}{2}$):

1. $\sin\left(\arcsin\frac{x}{a}\right) = \frac{x}{a}$,

2. $\cos\left(\arcsin\frac{x}{a}\right) = \sqrt{1 - [\sin(\arcsin(x/a))]^2} = \sqrt{1 - \left[\frac{x}{a}\right]^2}$

$$= \sqrt{\frac{a^2 - x^2}{a^2}} = \frac{\sqrt{a^2 - x^2}}{a},$$

3. $\tan\left(\arcsin\frac{x}{a}\right) = \dfrac{\sin\left(\arcsin\frac{x}{a}\right)}{\cos\left(\arcsin\frac{x}{a}\right)} = \dfrac{\frac{x}{a}}{\frac{\sqrt{a^2-x^2}}{a}} = \dfrac{x}{\sqrt{a^2-x^2}}$,

4. $\cot\left(\arcsin\frac{x}{a}\right) = \dfrac{1}{\tan\left(\arcsin\frac{x}{a}\right)} = \dfrac{1}{\frac{x}{\sqrt{a^2-x^2}}} = \dfrac{\sqrt{a^2-x^2}}{x}$,

5. $\csc\left(\arcsin\frac{x}{a}\right) = \dfrac{1}{\sin\left(\arcsin\frac{x}{a}\right)} = \dfrac{1}{\frac{x}{a}} = \dfrac{a}{x}$,

6. $\sec\left(\arcsin\frac{x}{a}\right) = \dfrac{1}{\cos\left(\arcsin\frac{x}{a}\right)} = \dfrac{1}{\frac{\sqrt{a^2-x^2}}{a}} = \dfrac{a}{\sqrt{a^2-x^2}}$.

In respect that

$$\sin\theta = \frac{x}{a},$$

it may be easier, however, to use the *reference triangle* approach for the same purposes, i.e., consider θ to be an *acute angle* in a *right triangle* with

$$\text{hypotenuse} = a, \quad \text{opposite leg} = x, \quad \text{and} \quad \text{adjacent leg} = \sqrt{a^2 - x^2}.$$

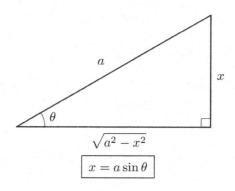

$$x = a\sin\theta$$

Example 6.2 (Integrals Containing $a^2 - x^2$).
For $a > 0$,

$$\int \sqrt{a^2 - x^2}\, dx \qquad\qquad \textit{substituting:} \quad \boxed{\begin{aligned} x &= a\sin\theta \\ dx &= a\cos\theta\, d\theta \\ \sqrt{a^2 - x^2} &= a\cos\theta \end{aligned}} ;$$

$$= \int a\cos\theta\, a\cos\theta\, d\theta = \int a^2\cos^2\theta\, d\theta$$

$$\text{by the \textit{power-reduction identity:}} \quad \boxed{\cos^2\theta = \frac{1+\cos 2\theta}{2}} ;$$

$$= \int \frac{a^2}{2}[1 + \cos 2\theta]\, d\theta \qquad\qquad \text{by the \textit{integration rules};}$$

$$= \frac{a^2}{2}\left[\int 1\, d\theta + \int \cos 2\theta\, d\theta\right] = \frac{a^2}{2}\left[\theta + \frac{1}{2}\sin 2\theta\right] + C$$

$$\text{by the \textit{double-angle identity:}} \quad \boxed{\frac{1}{2}\sin 2\theta = \sin\theta\cos\theta} ;$$

$$= \frac{a^2}{2}[\theta + \sin\theta\cos\theta] + C$$

$$\textit{substituting back} \text{ and using the \textit{reference triangle}:}$$

$$\boxed{\theta = \arcsin\frac{x}{a}, \ \sin\theta = \frac{x}{a}, \ \cos\theta = \frac{\sqrt{a^2-x^2}}{a}};$$

$$= \frac{a^2}{2}\left[\arcsin\frac{x}{a} + \frac{x}{a}\frac{\sqrt{a^2-x^2}}{a}\right] + C = \frac{x}{2}\sqrt{a^2-x^2} + \frac{a^2}{2}\arcsin\frac{x}{a} + C.$$

6.3.   Integrals Containing $x^2 + a^2$

For an integral containing the term $x^2 + a^2$, where $a > 0$, especially in the *radical* form $\sqrt{x^2 + a^2}$, we use the *tangent substitution*:

$$x = a\tan\theta, \ -\pi/2 < \theta < \pi/2.$$

Remark 6.2. The *range* for the new variable θ is purposely chosen to be $(-\pi/2, \pi/2)$, since

- on $(-\pi/2, \pi/2)$, the function $a\tan\theta$ is *one-to-one*, and thus, we can substitute back:

$$\theta = \arctan\frac{x}{a},$$

  and has the *largest range* of values: the *entire* real axis $(-\infty, \infty)$;
- also, on $(-\pi/2, \pi/2)$, $\sec\theta \geq 1$, which is instrumental for the *elimination* of the radical $\sqrt{x^2 + a^2}$ (see below).

By the *Pythagorean trigonometric identity*

$$\tan^2\theta + 1 = \sec^2\theta$$

(see Appendix C), we have:

$$x^2 + a^2 = a^2\tan^2\theta + a^2 = a^2(\tan^2\theta + 1) = a^2\sec^2\theta.$$

Hence, the substitution eliminates the *radical* $\sqrt{x^2 + a^2}$:

$$\sqrt{x^2 + a^2} = \sqrt{a^2\sec^2\theta} = |a||\sec\theta|$$

$$\text{considering that } a > 0 \text{ and } \sec\theta \geq 1 \text{ for } -\pi/2 < \theta < \pi/2$$

$$= a\sec\theta.$$

The substitution transforms the initial integral into a *trigonometric integral* relative to the new variable θ.

As the following example shows, *completing the square* may be a necessary preliminary step before making the required substitution.

Example 6.3 (Integrals Containing $x^2 + a^2$).

$$\int_1^{2\sqrt{3}-1} \frac{\sqrt{x^2 + 2x + 5}}{(x+1)^4}\, dx \qquad\qquad \text{completing the square:}$$

$$\boxed{x^2 + 2x + 5 = x^2 + 2x + 1 + 4 = (x+1)^2 + 2^2};$$

$$= \int_1^{2\sqrt{3}-1} \frac{\sqrt{(x+1)^2 + 2^2}}{(x+1)^4}\, dx$$

substituting and *changing* the integration limits:

$$\boxed{\begin{array}{l|l}
\begin{array}{l}
x + 1 = 2\tan\theta \\
dx = 2\sec^2\theta\, d\theta \\
\sqrt{(x+1)^2 + 2^2} = 2\sec\theta \\
\theta = \arctan\dfrac{x+1}{2}
\end{array}
&
\begin{array}{c|c}
x & \theta \\ \hline
1 & \pi/4 \\
2\sqrt{3}-1 & \pi/3
\end{array}
\end{array}};$$

$$= \int_{\pi/4}^{\pi/3} \frac{2\sec\theta}{16\tan^4\theta} 2\sec^2\theta\, d\theta = \int_{\pi/4}^{\pi/3} \frac{1}{4}\frac{\sec^3\theta}{\tan^4\theta}\, d\theta$$

by the *trigonometric identities:* $\boxed{\sec\theta = \dfrac{1}{\cos\theta}, \ \tan\theta = \dfrac{\sin\theta}{\cos\theta}};$

$$= \int_{\pi/4}^{\pi/3} \frac{1}{4}\frac{\dfrac{1}{\cos^3\theta}}{\dfrac{\sin^4\theta}{\cos^4\theta}}\, d\theta \qquad\qquad \text{simplifying;}$$

$$= \int_{\pi/4}^{\pi/3} \frac{1}{4}\frac{\cos\theta}{\sin^4\theta}\, d\theta \qquad\qquad \text{by the } \textit{integration rules;}$$

$$= \frac{1}{4}\int_{\pi/4}^{\pi/3} \frac{\cos\theta}{\sin^4\theta}\, d\theta$$

substituting again: $\boxed{u = \sin\theta, \ du = \cos\theta\, d\theta \quad \begin{array}{c|c} \theta & u \\ \hline \pi/4 & \sqrt{2}/2 \\ \pi/3 & \sqrt{3}/2 \end{array}};$

$$= \frac{1}{4}\int_{\sqrt{2}/2}^{\sqrt{3}/2} u^{-4}\, du \qquad\qquad \text{by the } \textit{Newton-Leibniz Formula;}$$

$$= \frac{1}{4}\cdot\frac{1}{(-3)}u^{-3}\Big|_{\sqrt{2}/2}^{\sqrt{3}/2} = -\frac{1}{12}\left[\frac{8}{3\sqrt{3}} - \frac{8}{2\sqrt{2}}\right] = -\frac{1}{12}\left[\frac{8\sqrt{3}}{9} - 2\sqrt{2}\right]$$

$$= \frac{\sqrt{2}}{6} - \frac{2\sqrt{3}}{27}.$$

What If the Integral Is Indefinite?

If the initial integral is *indefinite*, after having *substituted* and *evaluated* the integral relative to the new variable θ, we obtain an answer in terms of θ and/or some *trigonometric functions* of θ. Thus, when *substituting back*:

$$\theta = \arctan \frac{x}{a},$$

we are to encounter expressions containing $\arctan \frac{x}{a}$ inside trigonometric functions, which should be equivalently transformed into algebraic expressions in x not containing the trigonometric functions of $\arctan \frac{x}{a}$.

This can be done based on *cancellation* and *trigonometric identities* (taking into account that $-\frac{\pi}{2} < \arctan \frac{x}{a} < \frac{\pi}{2}$):

1. $\tan\left(\arctan \dfrac{x}{a}\right) = \dfrac{x}{a}$,

2. $\cot\left(\arctan \dfrac{x}{a}\right) = \dfrac{1}{\tan\left(\arctan \dfrac{x}{a}\right)} = \dfrac{1}{\dfrac{x}{a}} = \dfrac{a}{x}$,

3. $\sec\left(\arctan \dfrac{x}{a}\right) = \sqrt{[\tan(\arctan(x/a))]^2 + 1} = \sqrt{\left[\dfrac{x}{a}\right]^2 + 1}$

    $= \sqrt{\dfrac{x^2 + a^2}{a^2}} = \dfrac{\sqrt{x^2 + a^2}}{a}$,

4. $\cos\left(\arctan \dfrac{x}{a}\right) = \dfrac{1}{\sec\left(\arctan \dfrac{x}{a}\right)} = \dfrac{1}{\dfrac{\sqrt{x^2 + a^2}}{a}} = \dfrac{a}{\sqrt{x^2 + a^2}}$,

5. $\sin\left(\arctan \dfrac{x}{a}\right) = \tan\left(\arctan \dfrac{x}{a}\right)\cos\left(\arctan \dfrac{x}{a}\right) = \dfrac{x}{a} \cdot \dfrac{a}{\sqrt{x^2 + a^2}}$

    $= \dfrac{x}{\sqrt{x^2 + a^2}}$,

6. $\csc\left(\arctan \dfrac{x}{a}\right) = \dfrac{1}{\sin\left(\arctan \dfrac{x}{a}\right)} = \dfrac{1}{\dfrac{x}{\sqrt{x^2 + a^2}}} = \dfrac{\sqrt{x^2 + a^2}}{x}$.

In respect that

$$\tan \theta = \frac{x}{a},$$

it may be easier, however, to use the *reference triangle* approach for the same purposes, i.e., consider θ to be an *acute angle* in a *right triangle* with

$$\text{opposite leg} = x, \quad \text{adjacent leg} = a, \quad \text{and} \quad \text{hypotenuse} = \sqrt{x^2 + a^2}.$$

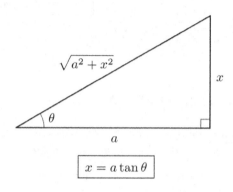

$$x = a \tan \theta$$

Example 6.4 (Integrals Containing $x^2 + a^2$).

$$\int \frac{x^3}{\sqrt{x^2 + 1}}\, dx = \int \frac{x^3}{\sqrt{x^2 + 1^2}}\, dx \qquad \text{substituting:} \quad \boxed{\begin{array}{l} x = \tan \theta \\ dx = \sec^2 \theta\, d\theta \\ \sqrt{x^2 + 1} = \sec \theta \end{array}};$$

$$= \int \frac{\tan^3 \theta}{\sec \theta} \sec^2 \theta\, d\theta = \int \tan^3 \theta \sec \theta\, d\theta$$

applying the appropriate *integration strategy*,
we *split* $\sec x \tan x$ off $\tan^3 \theta \sec \theta$;

$$= \int \tan^2 \theta (\sec \theta \tan \theta)\, d\theta$$

expressing $\tan^2 x$ in terms of $\sec x$ via $\boxed{\tan^2 \theta = \sec^2 \theta - 1}$;

$$= \int (\sec^2 \theta - 1)(\sec \theta \tan \theta)\, d\theta$$

substituting again: $\boxed{u = \sec \theta, \ du = \sec \theta \tan \theta\, d\theta}$;

$$= \int (u^2 - 1)\, du$$

by the *integration rules*;

$$= \int u^2\, du - \int 1\, du = \frac{u^3}{3} - u + C \qquad\qquad \textit{substituting back};$$

$$= \frac{\sec^3 \theta}{3} - \sec \theta + C$$

$$\textit{substituting back again and using the reference triangle:}$$

$$\boxed{\theta = \arctan x, \ \sec \theta = \sqrt{x^2 + 1}};$$

$$= \frac{(x^2+1)^{3/2}}{3} - \sqrt{x^2+1} + C = \frac{\sqrt{x^2+1}}{3}(x^2-2) + C.$$

6.4. Integrals Containing $x^2 - a^2$

For an integral containing the term $x^2 - a^2$, where $a > 0$, especially in the *radical* form $\sqrt{x^2 - a^2}$, we use the *secant substitution*:

$$x = a \sec \theta, \ 0 \le \theta < \pi/2 \text{ or } \pi/2 < \theta \le \pi.$$

Remark 6.3. The *range* for the new variable θ is purposely chosen to be $[0, \pi/2) \cup (\pi/2, \pi]$, since on $[0, \pi/2) \cup (\pi/2, \pi]$, the function $a \sec \theta$ is *one-to-one* and thus, we can substitute back:

$$\theta = \operatorname{arcsec} \frac{x}{a},$$

and has the *largest range* of values $(-\infty, a] \cup [a, \infty)$.

By the *Pythagorean trigonometric identity*

$$\sec^2 \theta - 1 = \tan^2 \theta$$

(see Appendix C), we have:

$$x^2 - a^2 = a^2 \sec^2 \theta - a^2 = a^2(\sec^2 \theta - 1) = a^2 \tan^2 \theta.$$

Hence, the substitution eliminates the *radical* $\sqrt{x^2 - a^2}$:

$$\sqrt{x^2 - a^2} = \sqrt{a^2 \tan^2 \theta} = |a||\tan \theta| \qquad\qquad \text{considering that } a > 0;$$

$$= \begin{cases} a \tan \theta & \text{for } 0 \le \theta < \pi/2 \ (x \ge a) \\ -a \tan \theta & \text{for } \pi/2 < \theta \le \pi \ (x \le -a) \end{cases} = \operatorname{sgn} x \cdot a \tan \theta,$$

$$\text{where } \operatorname{sgn} x = \begin{cases} 1 & \text{for } x \ge a \\ -1 & \text{for } x \le -a \end{cases}.$$

The substitution transforms the initial integral into a trigonometric integral relative to the new variable θ.

Example 6.5 (Integrals Containing $x^2 - a^2$).

$$\int_{\sqrt{2}}^{2} \frac{1}{(x^2 - 1)^{3/2}}\, dx = \int_{\sqrt{2}}^{2} \frac{1}{\left(\sqrt{x^2 - 1}\right)^3}\, dx$$

substituting:

$x = \sec\theta$		
$dx = \sec\theta\tan\theta\, d\theta$	x	θ
$\sqrt{x^2 - 1} = \tan\theta$	$\sqrt{2}$	$\pi/4$
$\theta = \operatorname{arcsec} x$	2	$\pi/3$

;

$$= \int_{\pi/4}^{\pi/3} \frac{1}{\tan^3 x} \sec x \tan x\, d\theta = \int_{\pi/4}^{\pi/3} \frac{\sec x}{\tan^2 x}\, d\theta$$

by the *trigonometric identities:* $\boxed{\sec\theta = \dfrac{1}{\cos\theta},\ \tan\theta = \dfrac{\sin\theta}{\cos\theta}}$;

$$= \int_{\pi/4}^{\pi/3} \frac{\dfrac{1}{\cos\theta}}{\dfrac{\sin^2\theta}{\cos^2\theta}}\, d\theta \qquad\qquad \text{simplifying and regrouping;}$$

$$= \int_{\pi/4}^{\pi/3} \frac{\cos\theta}{\sin^2\theta}\, d\theta = \int_{\pi/4}^{\pi/3} \frac{1}{\sin\theta} \frac{\cos\theta}{\sin\theta}\, d\theta$$

by the *trigonometric identities:* $\boxed{\dfrac{1}{\sin\theta} = \csc\theta,\ \dfrac{\cos\theta}{\sin\theta} = \cot\theta}$;

$$= \int_{\pi/4}^{\pi/3} \csc\theta\cot\theta\, d\theta \qquad\qquad \text{by the *Newton-Leibniz Formula*;}$$

$$= -\csc\theta\Big|_{\pi/4}^{\pi/3} = -[\csc(\pi/3) - \csc(\pi/4)] = -\left[\frac{2}{\sqrt{3}} - \sqrt{2}\right] = \sqrt{2} - \frac{2\sqrt{3}}{3}.$$

What If the Integral Is Indefinite?

If the initial integral is *indefinite*, after having *substituted* and *evaluated* the integral relative to the new variable θ, we obtain an answer in terms of θ and/or some *trigonometric functions* of θ. Thus, when *substituting back*:

$$\theta = \operatorname{arcsec} \frac{x}{a},$$

we are to encounter expressions containing $\operatorname{arcsec} \dfrac{x}{a}$ inside trigonometric functions, which should be equivalently transformed into algebraic expressions in x not containing the trigonometric functions of $\operatorname{arcsec} \dfrac{x}{a}$.

This can be done based on *cancellation* and *trigonometric identities* (taking into account that $0 \leq \operatorname{arcsec} \dfrac{x}{a} < \dfrac{\pi}{2}$ or $\dfrac{\pi}{2} < \operatorname{arcsec} \dfrac{x}{a} \leq \pi$):

1. $\sec\left(\operatorname{arcsec} \dfrac{x}{a}\right) = \dfrac{x}{a}$,

2. $\cos\left(\operatorname{arcsec} \dfrac{x}{a}\right) = \dfrac{1}{\sec\left(\operatorname{arcsec} \dfrac{x}{a}\right)} = \dfrac{1}{\dfrac{x}{a}} = \dfrac{a}{x}$,

3. $\sin\left(\operatorname{arcsec} \dfrac{x}{a}\right) = \sqrt{1 - \left[\cos\left(\operatorname{arcsec} \dfrac{x}{a}\right)\right]^2} = \sqrt{1 - \left[\dfrac{a}{x}\right]^2}$

$\qquad = \sqrt{\dfrac{x^2 - a^2}{x^2}} = \dfrac{\sqrt{x^2 - a^2}}{|x|}$,

4. $\csc\left(\operatorname{arcsec} \dfrac{x}{a}\right) = \dfrac{1}{\sin\left(\operatorname{arcsec} \dfrac{x}{a}\right)} = \dfrac{1}{\dfrac{\sqrt{x^2 - a^2}}{|x|}} = \dfrac{|x|}{\sqrt{x^2 - a^2}}$,

5. $\tan\left(\operatorname{arcsec} \dfrac{x}{a}\right) = \dfrac{\sin\left(\operatorname{arcsec} \dfrac{x}{a}\right)}{\cos\left(\operatorname{arcsec} \dfrac{x}{a}\right)} = \dfrac{\dfrac{\sqrt{x^2 - a^2}}{|x|}}{\dfrac{a}{x}} = \dfrac{x}{|x|} \dfrac{\sqrt{x^2 - a^2}}{a}$

$\qquad = \operatorname{sgn} x \dfrac{\sqrt{x^2 - a^2}}{a}$,

6. $\cot\left(\operatorname{arcsec} \dfrac{x}{a}\right) = \dfrac{1}{\tan\left(\operatorname{arcsec} \dfrac{x}{a}\right)} = \dfrac{1}{\operatorname{sgn} x \dfrac{\sqrt{x^2 - a^2}}{a}} = \operatorname{sgn} x \dfrac{a}{\sqrt{x^2 - a^2}}$.

In respect that

$$\sec \theta = \dfrac{x}{a},$$

it may be easier, however, to use the *reference triangle* approach for the same purposes, i.e., consider θ to be an *acute angle* in a *right triangle* with

hypotenuse $= x$,　adjacent leg $= a$,　and　opposite leg $= \sqrt{x^2 - a^2}$.

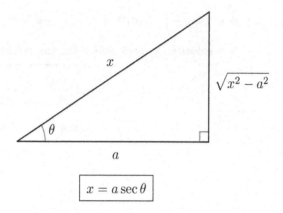

$$x = a \sec \theta$$

Remark 6.4. To include the case of $\dfrac{\pi}{2} < \arcsec \dfrac{x}{a} \le \pi$, when the *hypotenuse* $x = a \sec \theta$ of the reference triangle is *negative*, we need to *multiply* all trigonometric functions of $\arcsec \dfrac{x}{a}$ obtained from it, except $\sec\left(\arcsec \dfrac{x}{a}\right)$ and $\cos\left(\arcsec \dfrac{x}{a}\right)$, by $\operatorname{sgn} x$ taking into account that $|x| = \operatorname{sgn} x \cdot x$, $x = \operatorname{sgn} x |x|$, and $\operatorname{sgn}^2 x = 1$. For instance,

$$\sin\left(\arcsec \frac{x}{a}\right) = \operatorname{sgn} x \frac{\sqrt{x^2 - a^2}}{x} = \frac{\sqrt{x^2 - a^2}}{|x|}.$$

Example 6.6 (Integrals Containing $x^2 - a^2$).
For $x < -1/2$ or $x > 1/2$,

$$\int \frac{1}{x^2 \sqrt{4x^2 - 1}}\, dx = \int \frac{1}{2x^2 \sqrt{x^2 - 1/4}}\, dx = \int \frac{1}{2x^2 \sqrt{x^2 - (1/2)^2}}\, dx$$

$$\text{substituting:} \quad \boxed{\begin{aligned} x &= \frac{1}{2} \sec \theta \\ dx &= \frac{1}{2} \sec \theta \tan \theta\, d\theta \\ \sqrt{x^2 - 1/4} &= \operatorname{sgn} x \frac{1}{2} \tan \theta \end{aligned}} \quad ;$$

$$= \int \frac{1}{2\frac{1}{4} \sec^2 \theta \, \operatorname{sgn} x \frac{1}{2} \tan \theta} \, \frac{1}{2} \sec \theta \tan \theta \, d\theta \qquad \text{simplifying;}$$

$$= \int \frac{2 \operatorname{sgn} x}{\sec \theta} \, d\theta$$

by the *trigonometric identity* $\boxed{\dfrac{1}{\sec \theta} = \cos \theta}$ and the *integration rules*;

$$= \int 2\operatorname{sgn} x \cos\theta\, d\theta = 2\operatorname{sgn} x \int \cos\theta\, d\theta = 2\operatorname{sgn} x \sin\theta + C$$

substituting back and using the *reference triangle*:

$$\boxed{\sin\theta = \operatorname{sgn} x \frac{\sqrt{x^2 - 1/4}}{x}};$$

$$= 2\operatorname{sgn}^2 x \frac{\sqrt{x^2 - 1/4}}{x} + C \qquad\qquad \text{considering that } \operatorname{sgn}^2 x = 1;$$

$$= \frac{\sqrt{4x^2 - 1}}{x} + C.$$

6.5. Applications

Examples 6.7 (Applications).

1. Find the *area* of the *ellipse* $\dfrac{x^2}{a^2} + \dfrac{y^2}{b^2} = 1$ $(a, b > 0)$.

   Solution: The quarter of the ellipse in the *first quadrant* is explicitly defined as follows:

   $$y = \frac{b}{a}\sqrt{a^2 - x^2},\ 0 \le x \le a.$$

   Hence, by the symmetry,

   $$A = 4 \int_0^a \frac{b}{a}\sqrt{a^2 - x^2}\, dx$$

   substituting and *changing* the integration limits:

$x = a\sin\theta$	x	θ
$dx = a\cos\theta\, d\theta$		
$\sqrt{a^2 - x^2} = a\cos\theta$	0	0
$\theta = \arcsin\dfrac{x}{a}$	a	$\pi/2$

   $$= 4 \int_0^{\pi/2} \frac{b}{a} a^2 \cos^2\theta\, d\theta$$

   by the *power-reduction identity*: $\boxed{\cos^2\theta = \dfrac{1 + \cos 2\theta}{2}};$

   $$= 4 \int_0^{\pi/2} \frac{ab}{2}[1 + \cos 2\theta]\, d\theta \qquad\qquad \text{by the *integration rules*;}$$

$$= 2ab \left[\int_0^{\pi/2} 1 \, d\theta + \int_0^{\pi/2} \cos 2\theta \, d\theta \right]$$

by the *Newton-Leibniz Formula*;

$$= 2ab \left[\theta \Big|_0^{\pi/2} + \frac{1}{2} \sin 2\theta \Big|_0^{\pi/2} \right] = 2ab \left[\frac{\pi}{2} + \frac{1}{2}(\sin \pi - \sin 0) \right] = \pi ab \text{ un.}^2$$

(cf. Examples 6.2).

Remark 6.5. When $a = b = r$, we obtain the well-known *circle* (of radius r) *area* formula $A = \pi r^2$.

2. Find the *area* of the region bounded by the graph of $y = \dfrac{1}{(x^2 + 4)^2}$ and the x-axis on the interval $[-2, 2]$.

   Solution:

$$A = \int_{-2}^{2} \frac{1}{(x^2 + 4)^2} \, dx \qquad \text{by the } \textit{evenness} \text{ of the } \textit{integrand};$$

$$= 2 \int_0^2 \frac{1}{(x^2 + 4)^2} \, dx = 2 \int_0^2 \frac{1}{(x^2 + 2^2)^2} \, dx$$

substituting and *changing* the integration limits:

$x = 2 \tan \theta$		
$dx = 2 \sec^2 \theta \, d\theta$	x	θ
$x^2 + 2^2 = 4 \sec^2 \theta$	0	0
$\theta = \arctan \dfrac{x}{2}$	2	$\pi/4$

$$= 2 \int_0^{\pi/4} \frac{1}{4^2 \sec^4 \theta} 2 \sec^2 \theta \, d\theta \qquad \text{simplifying;}$$

$$= 2 \int_0^{\pi/4} \frac{1}{8 \sec^2 \theta} \, d\theta$$

by the *trigonometric identity*: $\boxed{\dfrac{1}{\sec \theta} = \cos \theta}$;

$$= 2 \int_0^{\pi/4} \frac{1}{8} \cos^2 \theta \, d\theta$$

by the *trigonometric identity*: $\boxed{\cos^2 \theta = \dfrac{1 + \cos 2\theta}{2}}$;

$$= 2 \int_0^{\pi/4} \frac{1}{8} \cos^2 \theta \, d\theta = 2 \int_0^{\pi/4} \frac{1}{16} [1 + \cos 2\theta] \, d\theta$$

by the *integration rules* and the *Newton-Leibniz Formula*;

$$= \frac{1}{8} \left[\int_0^{\pi/4} 1\, d\theta + \int_0^{\pi/4} \cos 2\theta\, d\theta \right] = \frac{1}{8} \left[\theta \Big|_0^{\pi/4} + \frac{1}{2} \sin 2\theta \Big|_0^{\pi/4} \right]$$

$$= \frac{1}{8} \left[\frac{\pi}{4} + \frac{1}{2} \right] = \frac{\pi}{32} + \frac{1}{16} \text{ un.}^2.$$

Exercise 6.1 (Alternate Approach).

Solve the last problem using *reduction formula* 11 of Appendix B.

6.6. Practice Problems

Evaluate the integrals.

1. $\displaystyle\int_0^1 \frac{1}{(4-x^2)^{3/2}}\, dx$

2. $\displaystyle\int_3^{3\sqrt{3}} \frac{x^2}{(x^2+9)^2}\, dx$

3. $\displaystyle\int_{2/\sqrt{3}}^2 \frac{1}{x^3\sqrt{x^2-1}}\, dx$

4. $\displaystyle\int_{-1}^0 \sqrt{x^2+2x+2}\, dx$

5. $\displaystyle\int \frac{x^5}{\sqrt{1-x^2}}\, dx$

6. $\displaystyle\int \frac{x^2}{\sqrt{x^2+2}}\, dx$

7. $\displaystyle\int \frac{x^2}{\sqrt{x^2-9}}\, dx$

8. $\displaystyle\int \frac{\sqrt{9-4x^2}}{x^2}\, dx$

9. $\displaystyle\int \frac{\sqrt{9x^2-16}}{x^3}\, dx$

10. $\displaystyle\int \frac{1}{(x-1)\sqrt{x^2-2}}\, dx$

11. $\displaystyle\int \frac{x^2+2x+1}{\sqrt{x^2+2x+10}}\, dx$

12. $\displaystyle\int \frac{1}{(x+1)^2\sqrt{x^2+2x-5}}\, dx$

13. $\displaystyle\int \frac{x^2+1}{x\sqrt{x^4+1}}\, dx$

14. Prove that, for $a > 0$,

$$\int \sqrt{x^2 \pm a^2}\, dx = \frac{x}{2}\sqrt{x^2 \pm a^2} \pm \frac{a^2}{2} \ln|x+\sqrt{x^2 \pm a^2}| + C \text{ (cf. Examples 4.4).}$$

Using the results of Examples 6.2 and the prior problem, evaluate the integrals.

15. $\displaystyle\int \sqrt{2+x-x^2}\, dx$

16. $\displaystyle\int x\sqrt{x^4+2x^2-1}\, dx$

Chapter 7

Integration of Rational Functions

7.1. Rational Functions

Definition 7.1 (Rational Function (Fraction)).
A *rational function* (or fraction) is a ratio

$$\frac{P(x)}{Q(x)},$$

of two polynomials

$$P(x) = a_m x^m + a_{m-1} x^{m-1} + \cdots + a_1 x + a_0$$

and

$$Q(x) = b_n x^n + b_{n-1} x^{n-1} + \cdots + b_1 x + b_0$$

with *degrees* $m, n = 0, 1, 2, \ldots$ and *real coefficients* $a_m, \ldots, a_0, b_n, \ldots, b_0$, $a_m, b_n \neq 0$.
A *fraction* is said to be in *reduced form* if the numerator and denominator have no common factors. By canceling all common factors, any fraction can be equivalently written in reduced form.
A *fraction* is said to be *proper* if the degree of the numerator is less than the degree of the denominator ($m < n$) and *improper* otherwise ($m \geq n$).

Remark 7.1. Any *polynomial* $P(x)$ is trivially a fraction with the denominator $Q(x) = 1$.

Examples 7.1 (Rational Functions).

1. $\dfrac{x^2 - 1}{x + 1} = \dfrac{(x - 1)(x + 1)}{x + 1}$ is an *improper* fraction *not* in reduced form.
   Canceling the common factor $x + 1$, we have:

   $$\frac{(x - 1)(x + 1)}{x + 1} = x - 1, \; x \neq -1.$$

2. $\dfrac{x^4 - 1}{x^2 + 2x} = \dfrac{(x-1)(x+1)(x^2+1)}{x(x+2)}$ is an *improper* fraction in reduced form (nothing to cancel).

3. $\dfrac{x-2}{x^2 - 3x + 2} = \dfrac{x-2}{(x-1)(x-2)}$ is a *proper* fraction *not* in reduced form. Canceling the common factor $x - 2$, we have:

$$\frac{x-2}{(x-1)(x-2)} = \frac{1}{x-1}, \ x \neq 2.$$

4. $\dfrac{x-1}{x^2 + x + 1}$ is a *proper* fraction in reduced form. Observe that the *quadratic polynomial* $x^2 + x + 1$ cannot be factored into two linear polynomials unlike $x^2 - 3x + 2 = (x-1)(x-2)$.

The integration of rational functions is executed via the *Partial Fraction Method* reducing such to the integration of so-called *partial fractions* of rather simple structure.

7.2.   Partial Fractions

Definition 7.2 (Partial Fractions).
A *partial fraction* is a proper fraction of one of the following *four types*:

Type 1: $\dfrac{A}{ax + b}$,

Type 2: $\dfrac{A}{(ax + b)^k}$, $k = 2, 3, \ldots$,

Type 3: $\dfrac{Ax + B}{ax^2 + bx + c}$,

Type 4: $\dfrac{Ax + B}{(ax^2 + bx + c)^l}$, $l = 2, 3, \ldots$,

where,

- a, b, and c are *real coefficients* with $a \neq 0$,
- A and B are *real coefficients* not simultaneously equal to 0 (i.e., $A^2 + B^2 \neq 0$), and
- $ax^2 + bx + c$ is an *irreducible quadratic polynomial*, i.e., such that cannot be further factored into two linear polynomials with real coefficients, which is the case *iff* $D = b^2 - 4ac < 0$.

Examples 7.2 (Partial Fractions).

1. $\dfrac{2}{3x - 1}$ is a *type 1* partial fraction ($A = 2$, $a = 3$, and $b = -1$).

2. $\dfrac{1}{(2x+5)^4}$ is a *type 2* partial fraction ($A = 1$, $a = 2$, and $b = 5$).

3. $\dfrac{x-1}{x^2+x+1}$ is a *type 3* partial fraction ($A = 1$, $B = -1$, $a = b = c = 1$).

4. $\dfrac{x-1}{(x^2+x+1)^2}$ is a *type 4* partial fraction ($A = 1$, $B = -1$, $a = b = c = 1$).

5. $\dfrac{2x+1}{x^2-3x+2}$ is a *not a partial fraction*, since $x^2 - 3x + 2$ is *not irre-* ducible: $x^2 - 3x + 2 = (x-1)(x-2)$.

7.2.1. *Integration of Type 1/Type 2 Partial Fractions*

A *type 1/type 2* partial fraction is integrated via the *trivial substitution* (see Sec. 3.1.4) or, which is the same, the *Useful Integration Formula* (Theorem 2.1) as follows:

$$\int \frac{A}{ax+b}\,dx = A\int \frac{1}{ax+b}\,dx = \frac{A}{a}\ln|ax+b| + C,$$

$$\int \frac{A}{(ax+b)^k}\,dx = A\int (ax+b)^{-k}\,dx = \frac{A}{a}\cdot \frac{1}{-k+1}(ax+b)^{-k+1} + C,$$

$k = 2, 3, \ldots .$

Examples 7.3 (Integration of Type 1/Type 2 Partial Fractions).

1. $\displaystyle\int \frac{2}{3x-1}\,dx = 2\int \frac{1}{3x-1}\,dx = \frac{2}{3}\ln|3x-1| + C.$

2. $\displaystyle\int \frac{1}{(2x+5)^4}\,dx = \int (2x+5)^{-4}\,dx = \frac{1}{2}\cdot \frac{1}{-3}(2x+5)^{-3} + C$

   $= -\dfrac{1}{6}\dfrac{1}{(2x+5)^3} + C.$

7.2.2. *Integration of Type 3 Partial Fractions*

Generally, a *type 3* partial fraction

$$\int \frac{Ax+B}{ax^2+bx+c}\,dx$$

is integrated via two procedures: *substitution* and *completing the square* (see Sec. 2.2.7).

Example 7.4 (Integration of Type 3 Partial Fractions).

$$\int \frac{x-1}{x^2+x+1}\,dx$$

since $[x^2+x+1]' = 2x+1$, multiplying and dividing by 2;

$$= \frac{1}{2}\int \frac{2x-2}{x^2+x+1}\,dx \qquad\qquad \text{adding and subtracting 1;}$$

$$= \frac{1}{2}\int \frac{2x+1-3}{x^2+x+1}\,dx \qquad\qquad \text{by the \textit{integration rules};}$$

$$= \frac{1}{2}\int \frac{2x+1}{x^2+x+1}\,dx - \frac{3}{2}\int \frac{1}{x^2+x+1}\,dx$$

substituting for the first integral: $\boxed{u = x^2+x+1,\ du = (2x+1)dx}$

and *completing the square* for the second one:

$$\boxed{x^2+x+1 = x^2+2(1/2)x+1/4-1/4+1 = (x+1/2)^2+(\sqrt{3}/2)^2}\,;$$

$$= \frac{1}{2}\int \frac{1}{u}\,du - \frac{3}{2}\int \frac{1}{(x+1/2)^2+(\sqrt{3}/2)^2}\,dx$$

$$= \frac{1}{2}\ln|u| - \frac{3}{2}\frac{1}{\sqrt{3}/2}\arctan\frac{x+1/2}{\sqrt{3}/2} + C \qquad \text{\textit{substituting back};}$$

$$= \frac{1}{2}\ln|x^2+x+1| - \sqrt{3}\arctan\frac{2x+1}{\sqrt{3}} + C \qquad \text{since } x^2+x+1 > 0;$$

$$= \frac{1}{2}\ln(x^2+x+1) - \sqrt{3}\arctan\frac{2x+1}{\sqrt{3}} + C.$$

Observations

When integrating a *type 3* partial fraction,

- In certain cases, only one of the two procedures is involved: *substitution*, as is the case for

$$\int \frac{2x+1}{x^2+x+1}\,dx = \int \frac{1}{u}\,du \text{ with } \boxed{u = x^2+x+1,\ du = (2x+1)dx},$$

  or *completing the square*, as is the case for

$$\int \frac{1}{x^2+x+1}\,dx = \int \frac{1}{(x+1/2)^2+(\sqrt{3}/2)^2}\,dx.$$

- If *substitution* is involved, it is always of the form

$$\boxed{u = ax^2+bx+c,\ du = (2ax+b)dx}$$

  and leads to the integral of the form

$$\int \frac{1}{u}\,du = \ln|u| + C.$$

- If *completing the square* is involved, due to the fact that the *quadratic polynomial* $ax^2 + bx + c$ is *irreducible*, it always leads to the integral of the form (up to a constant factor)

$$\int \frac{1}{(x+h)^2 + d^2}\, dx = \frac{1}{d} \arctan \frac{x+h}{d} + C.$$

7.2.3. Integration of Type 4 Partial Fractions

Just as in the case of a *type 3* partial fraction, when integrating a *type 4* partial fraction

$$\int \frac{Ax + B}{(ax^2 + bx + c)^l}\, dx, \quad l = 2, 3, \ldots,$$

we apply *substitution* and *completing the square*. However, when the latter is involved, we also use *reduction formula* 11 of Appendix B (see Examples 4.6), i.e., *partial integration* implicitly.

Example 7.5 (Integration of Type 4 Partial Fractions).

$$\int \frac{x-1}{(x^2 + x + 1)^2}\, dx \qquad \text{in the same manner as in the prior example;}$$

$$= \frac{1}{2} \int \frac{2x+1}{(x^2+x+1)^2}\, dx - \frac{3}{2} \int \frac{1}{(x^2+x+1)^2}\, dx$$

substituting for the first integral: $\boxed{u = x^2 + x + 1, \ du = (2x+1)dx}$

and *completing the square* for the second one (see the prior example);

$$= \frac{1}{2} \int u^{-2}\, du - \frac{3}{2} \int \frac{1}{\left[(x+1/2)^2 + (\sqrt{3}/2)^2\right]^2}\, dx$$

integrating and using *reduction formula* 11 of Appendix B

with $x + 1/2$ for x, $a = \sqrt{3}/2$, and $n = 2$ for the second integral;

$$= \frac{1}{2}\left[-\frac{1}{u}\right] - \frac{3}{2}\frac{1}{2(\sqrt{3}/2)^2}\left[\frac{x+1/2}{(x+1/2)^2 + (\sqrt{3}/2)^2}\right.$$

$$+ \left. \int \frac{1}{(x+1/2)^2 + (\sqrt{3}/2)^2}\, dx\right]$$

substituting back, integrating, and simplifying;

$$= -\frac{1}{2}\frac{1}{x^2+x+1} - \frac{x+1/2}{x^2+x+1} - \frac{2}{\sqrt{3}} \arctan \frac{2x+1}{\sqrt{3}} + C$$

$$= -\frac{x+1}{x^2+x+1} - \frac{2}{\sqrt{3}} \arctan \frac{2x+1}{\sqrt{3}} + C.$$

Observations

When integrating a *type 4* partial fraction,

- In certain cases, only one of the two procedures is involved: *substitution*, as is the case for

$$\int \frac{2x+1}{(x^2+x+1)^2}\, dx = \int u^{-2}\, du$$

with $\boxed{u = x^2 + x + 1,\ du = (2x+1)dx}$,

or *completing the square*, as is the case for

$$\int \frac{1}{(x^2+x+1)^2}\, dx = \int \frac{1}{\left[(x+1/2)^2 + (\sqrt{3}/2)^2\right]^2}\, dx.$$

- If *substitution* is involved, it is always of the form

$$\boxed{u = ax^2 + bx + c,\ du = (2ax+b)dx}$$

and leads to the integral of the form

$$\int u^{-l}\, du = \frac{1}{-l+1}u^{-l+1} + C, \quad l = 2, 3, \ldots.$$

- If *completing the square* is involved, due to the fact that the *quadratic polynomial* $ax^2 + bx + c$ is *irreducible*, it always leads to the integral of the form (up to a constant factor)

$$\int \frac{1}{\left[(x+h)^2 + d^2\right]^l}\, dx, \quad l = 2, 3, \ldots,$$

to which we apply *reduction formula* 11 of Appendix B $l-1$ times.

7.3.   Partial Fraction Decomposition

Observe that, by the *Fundamental Theorem of Algebra* (see, e.g., [1, 6]), any polynomial $Q(x)$ with real coefficient of degree $n = 1, 2, \ldots$ can be factored into factors of the form

$$(ax+b)^k \quad \text{or} \quad (ax^2 + bx + c)^l,$$

where k and l are *nonnegative integer exponents*, a, b, and c are *real coefficients* with $a \neq 0$ and $ax^2 + bx + c$ is an *irreducible* quadratic polynomial.

Examples 7.6 (Factoring Polynomials).

  1. $2x^2 - x - 1 = (2x+1)(x-1)$.

2. $x^3 - 8 = (x - 2)(x^2 + 2x + 4)$.

3. $x^4 + x^3 - x^2 - 5x + 4 = (x-1)(x^3 + 2x^2 + x - 4) = (x-1)^2(x^2 + 3x + 4)$.

4. $x^6 + 1 = (x^2 + 1)(x^4 - x^2 + 1) = (x^2 + 1)(x^4 + 2x^2 + 1 - 3x^2)$

$$= (x^2 + 1)\left[(x^2 + 1)^2 - (\sqrt{3}x)^2\right] = (x^2 + 1)(x^2 - \sqrt{3}x + 1)(x^2 + \sqrt{3}x + 1).$$

Theorem 7.1 (Partial Fraction Decomposition).

Suppose that $\dfrac{P(x)}{Q(x)}$ *is a* proper fraction, *with the denominator* $Q(x)$ *factored into factors of the form*

$$(ax + b)^k \quad or \quad (ax^2 + bx + c)^l,$$

where k and l are nonnegative integer exponents, a, b, and c are real coefficients with $a \neq 0$ and $ax^2 + bx + c$ is an irreducible quadratic polynomial. Then $\dfrac{P(x)}{Q(x)}$ *is decomposed into a sum of partial fractions in the following fashion:*

- *Each factor of the form $(ax + b)^k$ ($k = 1, 2, \dots$) contributes the sum of k type 1/type 2 partial fractions:*

$$\frac{A_1}{ax + b} + \frac{A_2}{(ax + b)^2} + \cdots + \frac{A_k}{(ax + b)^k}.$$

- *Each factor of the form $(ax^2 + bx + c)^l$ ($l = 1, 2, \dots$) into $Q(x)$ contributes the sum of l type 3/type 4 partial fractions:*

$$\frac{A_1 x + B_1}{ax^2 + bx + c} + \frac{A_2 x + B_2}{(ax^2 + bx + c)^2} + \cdots + \frac{A_l x + B_l}{(ax^2 + bx + c)^l}.$$

- *The partial fraction decomposition is obtained by adding the contributions of all the factors of the denominator.*

Algorithm of Partial Fraction Decomposition

To decompose a *proper fraction* $\dfrac{P(x)}{Q(x)}$ into *partial fractions,*

1. *Factor* the denominator $Q(x)$ completely.
2. *Set up* the appropriate *form* of the *partial fraction decomposition* by Theorem 7.1.
3. *Clear* the denominators multiplying through by the *common denominator* $Q(x)$.

4. *Determine* the values of the unknown coefficients by the *Method of Un-determined Coefficients*, i.e., by matching the like powers and equating the coefficients of the *polynomials* in the left and right sides of the equality at each one.
5. *Substitute* the found values for the corresponding coefficients into the *form* of *step 2* to write the desired partial fraction decomposition.

Example 7.7 (Partial Fraction Decomposition).

Decompose the *proper fraction*

$$\frac{5x^3 - 2x^2 + 10x - 5}{x^4 + x^3 - x^2 - 5x + 4}$$

into *partial fractions*.

Solution:

1. *Factor* the denominator:

$$x^4 + x^3 - x^2 - 5x + 4 = (x-1)^2(x^2 + 3x + 4)$$

(see Examples 7.6).

2. *Set up* the appropriate form of the *partial fraction decomposition*:

$$\frac{5x^3 - 2x^2 + 10x - 5}{(x-1)^2(x^2 + 3x + 4)} = \frac{A}{x-1} + \frac{B}{(x-1)^2} + \frac{Cx + D}{x^2 + 3x + 4}$$

with the coefficients A, B, C, and D to be determined.

3. *Clear* the denominators multiplying both sides by the *common denominator* $(x-1)^2(x^2 + 3x + 4)$:

$$5x^3 - 2x^2 + 10x - 5 = A(x-1)(x^2+3x+4) + B(x^2+3x+4) + (Cx+D)(x-1)^2$$

4. *Determine* the values of the unknown coefficients by the *Method of Un-determined Coefficients*.
Matching the like powers x^3, x^2, x^1, and x^0 and equating the coefficients of the *polynomials* in the left and right sides of the equality at each one, we obtain the following *linear system* of four equations in four unknowns:

$$
\begin{array}{llll}
\text{at } x^3: & A & + C & = 5 \\
\text{at } x^2: & 2A + B & - 2C + D & = -2 \\
\text{at } x^1: & A + 3B & + C - 2D & = 10 \\
\text{at } x^0: & -4A + 4B & + D & = -5
\end{array}
$$

(*verify*), solving which via the *elimination method*:

$$A \qquad + C \qquad = 5$$
$$2A + B - 2C + D = -2$$
$$A + 3B + C - 2D = 10$$
$$-4A + 4B \qquad + D = -5$$

$$\xrightarrow[E_4+4E_1\to E_4]{E_2+(-2)E_1\to E_2,\; E_3+(-1)E_1\to E_3}$$

$$A \qquad + C \qquad = 5$$
$$B - 4C + D = -12$$
$$3B \qquad - 2D = 5$$
$$4B + 4C + D = 15$$

$$\xrightarrow[E_4+(-4)E_2\to E_4]{E_3+(-3)E_2\to E_3}$$

$$A \qquad + C \qquad = 5$$
$$B - 4C + D = -12$$
$$12C - 5D = 41$$
$$20C - 3D = 63$$

$$\xrightarrow{3E_4+(-5)E_3\to E_4}$$

$$A \qquad + C \qquad = 5$$
$$B - 4C + D = -12$$
$$12C - 5D = 41$$
$$16D = -16$$

solving starting with the last equation and *back substituting*;

$$A \qquad = \qquad 5 - 3 = 2$$
$$B \qquad = -12 + 4 \cdot 3 - (-1) = 1$$
$$C \qquad = \quad [41 + 5(-1)]/12 = 3$$
$$D = \qquad -1$$

we arrive at the *unique solution:* $A = 2$, $B = 1$, $C = 3$, $D = -1$.

5. *Substituting* the found values for the corresponding coefficients into the *form* of step 2, we obtain the partial fraction decomposition

$$\frac{5x^3 - 2x^2 + 10x - 5}{(x-1)^2(x^2+3x+4)} = \frac{2}{x-1} + \frac{1}{(x-1)^2} + \frac{3x-1}{x^2+3x+4}.$$

7.4. Partial Fraction Method

The *Partial Fraction Method* is used for the evaluation of integrals of rational functions.

Given an integral of a rational function

$$\int \frac{P(x)}{Q(x)} \, dx,$$

the method is executed as follows:

- If the fraction $\dfrac{P(x)}{Q(x)}$ is *proper*,

  1. *Decompose* into partial fractions.
  2. *Integrate.*

- If the fraction $\dfrac{P(x)}{Q(x)}$ is *improper*,

  1. *Divide:* $\dfrac{P(x)}{Q(x)} = S(x) + \dfrac{R(x)}{Q(x)}$, where the *quotient* $S(x)$ and the *remainder* $R(x)$ are polynomials, the degree of $R(x)$ being *less* than the degree of $Q(x)$.
  2. *Decompose* the *proper fraction* $\dfrac{R(x)}{Q(x)}$ into partial fractions.
  3. *Integrate.*

Example 7.8 (Partial Fraction Method).
Evaluate the integral

$$\int \frac{2x^5 + 3x^4 + 4x^3 - 13x^2 + 13x - 1}{x^4 + x^3 - x^2 - 5x + 4} \, dx.$$

Solution:

1. Since the fraction is *improper*, we use *long division* first:

$$\frac{2x^5 + 3x^4 + 4x^3 - 13x^2 + 13x - 1}{x^4 + x^3 - x^2 - 5x + 4} = 2x + 1 + \frac{5x^3 - 2x^2 + 10x - 5}{x^4 + x^3 - x^2 - 5x + 4}$$

   (*verify*).

2. *Decomposing* the *proper fraction*

$$\frac{5x^3 - 2x^2 + 10x - 5}{x^4 + x^3 - x^2 - 5x + 4}$$

   into *partial fractions* (see Example 7.7), we have:

$$\frac{5x^3 - 2x^2 + 10x - 5}{x^4 + x^3 - x^2 - 5x + 4} = \frac{2}{x - 1} + \frac{1}{(x - 1)^2} + \frac{3x - 1}{x^2 + 3x + 4}.$$

3. Now, we *integrate*:

$$\int \frac{2x^5 + 3x^4 + 4x^3 - 13x^2 + 13x - 1}{x^4 + x^3 - x^2 - 5x + 4}\, dx$$

$$= \int \left[2x + 1 + \frac{2}{x-1} + \frac{1}{(x-1)^2} + \frac{3x-1}{x^2+3x+4} \right] dx$$

by the *integration rules*;

$$= \int 2x\, dx + \int 1\, dx + 2\int \frac{1}{x-1}\, dx + \int (x-1)^{-2}\, dx + \int \frac{3x-1}{x^2+3x+4}\, dx$$

it is easier to see what adjustments are required for the forthcoming substitution when the latter integral is split as follows:

$$= x^2 + x + 2\ln|x-1| - \frac{1}{x-1} + 3\int \frac{x}{x^2+3x+4}\, dx - \int \frac{1}{x^2+3x+4}\, dx$$

since $[x^2+3x+4]' = 2x+3$,

introducing the *missing constants* 2 and 3 for the first integral;

$$= x^2 + x + 2\ln|x-1| - \frac{1}{x-1} + \frac{3}{2}\int \frac{2x+3-3}{x^2+3x+4}\, dx - \int \frac{1}{x^2+3x+4}\, dx$$

by the *integration rules*;

$$= x^2 + x + 2\ln|x-1| - \frac{1}{x-1} + \frac{3}{2}\int \frac{2x+3}{x^2+3x+4}\, dx - \frac{9}{2}\int \frac{1}{x^2+3x+4}\, dx$$

$$- \int \frac{1}{x^2+3x+4}\, dx = x^2 + x + 2\ln|x-1| - \frac{1}{x-1} + \frac{3}{2}\int \frac{2x+3}{x^2+3x+4}\, dx$$

$$- \frac{11}{2}\int \frac{1}{x^2+3x+4}\, dx$$

substituting for the first integral:

$$\boxed{u = x^2 + 3x + 4, \quad du = (2x+3)dx}$$

and *completing the square* for the second one:

$$\boxed{x^2 + 3x + 4 = x^2 + 2(3/2)x + 9/4 - 9/4 + 4 = (x+3/2)^2 + (\sqrt{7}/2)^2}\,;$$

$$= x^2 + x + 2\ln|x-1| - \frac{1}{x-1} + \frac{3}{2}\int \frac{1}{u}\, du - \frac{11}{2}\int \frac{1}{(x+3/2)^2 + (\sqrt{7}/2)^2}\, dx$$

$$= x^2 + x + 2\ln|x-1| - \frac{1}{x-1} + \frac{3}{2}\ln|u| - \frac{11}{2}\frac{1}{\sqrt{7}/2}\arctan\frac{x+3/2}{\sqrt{7}/2} + C$$

substituting back and *simplifying*;

$$= x^2 + x + 2\ln|x-1| - \frac{1}{x-1} + \frac{3}{2}\ln|x^2+3x+4| - \frac{11}{\sqrt{7}}\arctan\frac{2x+3}{\sqrt{7}} + C$$

$$\text{since } x^2 + 3x + 4 > 0;$$

$$= x^2 + x + 2\ln|x-1| - \frac{1}{x-1} + \frac{3}{2}\ln(x^2+3x+4) - \frac{11}{\sqrt{7}}\arctan\frac{2x+3}{\sqrt{7}} + C.$$

Let us consider one more example, in which we are to deal with a *type 4* partial fraction.

Example 7.9 (Partial Fraction Method).
Evaluate the integral

$$\int \frac{x^3+1}{x(x^2+x+1)^2}\, dx.$$

Solution:

1. Observe that the integrand is a *proper fraction*, whose denominator is *completely factored*. Its *partial fraction decomposition* is of the form:

$$\frac{x^3+1}{x(x^2+x+1)^2} = \frac{A}{x} + \frac{Bx+C}{x^2+x+1} + \frac{Dx+E}{(x^2+x+1)^2}.$$

Multiplying through by the *common denominator* $x(x^2+x+1)^2$, we have:

$$x^3 + 1 = A(x^2+x+1)^2 + (Bx+C)x(x^2+x+1) + (Dx+E)x.$$

Considering that

$$(x^2+x+1)^2 = x^4 + 2x^3 + 3x^2 + 2x + 1$$

and

$$x(x^2+x+1) = x^3 + x^2 + x,$$

comparing the corresponding coefficients, we obtain the following 5×5 *linear system*:

$$
\begin{array}{llll}
\text{at } x^4: & A + B & & = 0 \\
\text{at } x^3: & 2A + B + C & & = 1 \\
\text{at } x^2: & 3A + B + C + D & & = 0 \\
\text{at } x^1: & 2A \quad + C \quad + E & = 0 \\
\text{at } x^4: & A & & = 1
\end{array}
$$

- From the last equation, $A = 1$.
- Substituting the found value for A into the *first* equation and solving it for B, we find $B = -1$.

- Substituting the found values for A and B into the *second* equation and solving it for C, we find $C = 1 - 2 \cdot 1 - (-1) = 0$.
- Substituting the found values for A, B, and C into the *third* equation and solving it for D, we find $D = -3 \cdot 1 - (-1) - 0 = -2$.
- Substituting the found values for A and C into the *fourth* equation and solving it for E, we find $E = -2 \cdot 1 - 0 = -2$.

Thus, the desired *partial fraction decomposition* is

$$\frac{x^3 + 1}{x(x^2 + x + 1)^2} = \frac{1}{x} + \frac{(-1)x}{x^2 + x + 1} + \frac{(-2)x + (-2)}{(x^2 + x + 1)^2}.$$

2. Now, we *integrate*:

$$\int \frac{x^3 + 1}{x(x^2 + x + 1)^2}\, dx = \int \left[\frac{1}{x} + \frac{(-1)x}{x^2 + x + 1} + \frac{(-2)x + (-2)}{(x^2 + x + 1)^2}\right] dx$$

by the *integration rules*;

$$= \int \frac{1}{x}\, dx - \int \frac{x}{x^2 + x + 1}\, dx - \int \frac{2x + 2}{(x^2 + x + 1)^2}\, dx$$

since $[x^2 + x + 1]' = 2x + 1$,

introducing the *missing constants* 2 and 1

for the second integral and expressing 2 as $1 + 1$ in the third one;

$$= \ln|x| - \frac{1}{2}\int \frac{2x + 1 - 1}{x^2 + x + 1}\, dx - \int \frac{2x + 1 + 1}{(x^2 + x + 1)^2}\, dx$$

by the *integration rules*;

$$= \ln|x| - \frac{1}{2}\int \frac{2x + 1}{x^2 + x + 1}\, dx + \frac{1}{2}\int \frac{1}{x^2 + x + 1}\, dx - \int \frac{2x + 1}{(x^2 + x + 1)^2}\, dx$$
$$- \int \frac{1}{(x^2 + x + 1)^2}\, dx$$

substituting for the first and third integrals:

$$\boxed{u = x^2 + x + 1, \ du = (2x + 1)\, dx}$$

and *completing the square* for the second and fourth ones:

$$\boxed{x^2 + x + 1 = x^2 + 2(1/2)x + 1/4 - 1/4 + 1 = (x + 1/2)^2 + (\sqrt{3}/2)^2};$$
$$= \ln|x| - \frac{1}{2}\int \frac{1}{u}\, du + \frac{1}{2}\int \frac{1}{(x + 1/2)^2 + (\sqrt{3}/2)^2}\, dx - \int u^{-2}\, du$$
$$- \int \frac{1}{\left[(x + 1/2)^2 + (\sqrt{3}/2)^2\right]^2}\, dx$$

integrating and using *reduction formula* 11 of Appendix B
with $x + 1/2$ for x, $a = \sqrt{3}/2$, and $n = 2$ for the fourth integral;

$$= \ln|x| - \frac{1}{2}\ln|u| + \frac{1}{2}\frac{1}{\sqrt{3}/2}\arctan\frac{x+1/2}{\sqrt{3}/2} + \frac{1}{u}$$

$$- \frac{1}{2(\sqrt{3}/2)^2}\left[\frac{x+1/2}{(x+1/2)^2 + (\sqrt{3}/2)^2} + \int \frac{1}{(x+1/2)^2 + (\sqrt{3}/2)^2}\,dx\right]$$

substituting back and simplifying;

$$= \ln|x| - \frac{1}{2}\ln|x^2 + x + 1| + \frac{1}{\sqrt{3}}\arctan\frac{2x+1}{\sqrt{3}} + \frac{1}{x^2+x+1}$$

$$- \frac{2}{3}\left[\frac{x+1/2}{x^2+x+1} + \frac{2}{\sqrt{3}}\arctan\frac{2x+1}{\sqrt{3}}\right] + C$$

$$= \ln|x| - \frac{1}{2}\ln|x^2 + x + 1| - \frac{2}{3}\frac{x-1}{x^2+x+1} - \frac{1}{3\sqrt{3}}\arctan\frac{2x+1}{\sqrt{3}} + C$$

since $x^2 + x + 1 > 0$;

$$= \ln|x| - \frac{1}{2}\ln(x^2 + x + 1) - \frac{2}{3}\frac{x-1}{x^2+x+1} - \frac{1}{3\sqrt{3}}\arctan\frac{2x+1}{\sqrt{3}} + C.$$

7.5. Applications

Examples 7.10 (Applications).

1. Find the *volume* of the solid obtained by rotating the region bounded
by $y = \dfrac{1}{\sqrt{x(5-x)}}$ and $y = \dfrac{1}{2}$ on the interval $[1, 4]$ about the *x-axis*.

Solution: Considering that

$$0 \le \frac{1}{\sqrt{x(5-x)}} \le \frac{1}{2} \quad \text{on } [1, 4],$$

by the *washer method*,

$$V = \int_1^4 \pi\left[\left(\frac{1}{2}\right)^2 - \left(\frac{1}{\sqrt{x(5-x)}}\right)^2\right]\,dx \qquad \text{by the *integration rules*;}$$

$$= \pi\left[\frac{1}{4}\int_1^4 1\,dx - \int_1^4 \frac{1}{x(5-x)}\,dx\right].$$

To decompose the proper fraction $\dfrac{1}{x(5-x)}$, let us use the following shortcut, similar to how it is done when *transforming products into sums* (see Sec. 2.2.5):

$$\frac{1}{x(5-x)} = \frac{1}{5}\left[\frac{1}{x} + \frac{1}{5-x}\right].$$

Hence,

$$V = \pi\left[\frac{1}{4}\int_1^4 1\,dx - \int_1^4 \frac{1}{x(5-x)}\,dx\right] \qquad \text{by *integration rules*;}$$

$$= \pi\left[\frac{1}{4}\int_1^4 1\,dx - \frac{1}{5}\left(\int_1^4 \frac{1}{x}\,dx + \int_1^4 \frac{1}{5-x}\,dx\right)\right]$$

$$\text{by the *Newton-Leibniz Formula*;}$$

$$= \pi\left[\frac{1}{4}x\Big|_1^4 - \frac{1}{5}\left(\ln|x|\Big|_1^4 - \ln|5-x|\Big|_1^4\right)\right]$$

$$= \pi\left[\frac{1}{4}(4-1) - \frac{1}{5}(\ln 4 - \ln 1 - [\ln 1 - \ln 4])\right]$$

$$\text{since } \ln 1 = 0 \text{ and } \ln 4 = 2\ln 2;$$

$$= \pi\left[\frac{3}{4} - \frac{2\ln 4}{5}\right] = \pi\left[\frac{3}{4} - \frac{4\ln 2}{5}\right] \text{ un.}^3.$$

2. Find the *volume* of the solid obtained by rotating the region bounded by $y = \dfrac{1}{x^2(x-2)}$ and *x-axis* on the interval $[3,4]$ about *y-axis*.

Solution: Since the rotation axis is perpendicular to the axis of definition, by the *shell method*,

$$V = \int_3^4 2\pi x \frac{1}{x^2(x-2)}\,dx \qquad \text{by the *integration rules*;}$$

$$= 2\pi \int_3^4 \frac{1}{x(x-2)}\,dx$$

$$\text{*decomposing into partial fractions* via the shortcut:}$$

$$\boxed{\frac{1}{x(x-2)} = \frac{1}{2}\left[\frac{1}{x-2} - \frac{1}{x}\right]} \quad \text{(cf. Sec. 2.2.5);}$$

$$= 2\pi \int_3^4 \frac{1}{2}\left[\frac{1}{x-2} - \frac{1}{x}\right]dx \qquad \text{by the *integration rules*;}$$

$$= 2\pi \frac{1}{2} \left[\int_3^4 \frac{1}{x-2} \, dx - \int_3^4 \frac{1}{x} \, dx \right] \quad \text{by the } \textit{Newton-Leibniz Formula;}$$

$$= \pi \left[\ln|x-2| \Big|_3^4 - \ln|x| \Big|_3^4 \right] = \pi \left[\ln 2 - \ln 1 - (\ln 4 - \ln 3) \right]$$

$$\text{since } \ln 1 = 0 \text{ and } \ln 4 = 2\ln 2;$$

$$= \pi \left[\ln 3 - \ln 2 \right] = \pi \ln \frac{3}{2} \text{ un.}^3.$$

7.6.   Practice Problems

Evaluate the integrals.

1. $\displaystyle\int \frac{2x+3}{(x+2)(x-5)} \, dx$

2. $\displaystyle\int \frac{x}{x^2+x-2} \, dx$

3. $\displaystyle\int \frac{x}{(x^2+1)(x^2+4)} \, dx$

4. $\displaystyle\int \frac{1}{x^3+1} \, dx$

5. $\displaystyle\int \frac{x^3+1}{x^3-5x^2+6x} \, dx$

6. $\displaystyle\int \left(\frac{x}{x^2+3x+2} \right)^2 dx$

7. $\displaystyle\int \frac{1}{(x^4-1)^2} \, dx$

8. $\displaystyle\int \frac{x^2}{x^2+2x+1} \, dx$

9. $\displaystyle\int \frac{x^3}{(x-1)^{100}} \, dx$

10. $\displaystyle\int \frac{x^{11}}{x^8+3x^4+2} \, dx$

11. $\displaystyle\int \frac{1}{x(x^{10}-1)^2} \, dx$

12. $\displaystyle\int \frac{1-x^7}{x(x^7+1)} \, dx$

13. $\displaystyle\int \frac{x^4+1}{x(x^4+5)(x^5+5x+1)} \, dx$

14. $\displaystyle\int \frac{x^2+1}{x^4+3x^2+1} \, dx$

Hint: In problems 8–14, it is recommended to avoid the direct application of the *Partial Fraction Method*.

Chapter 8

Rationalizing Substitutions

Definition 8.1 (Rationalizing Substitution).
A *rationalizing substitution* is a substitution transforming an integral of an *irrational* function, which may contain *radicals, exponentials, logarithms*, or *trigonometric functions*, into an integral of a *rational function*.
In this case, we say that the integral is *rationalized*.

Here, we consider several integration scenarios implemented via *rationalizing substitutions*.

8.1. Integrals with Radicals

Integrals with *radicals* may be rationalized via an appropriate substitutions.

8.1.1. Integrals of the Form $\int R\left(x, \sqrt[n]{\dfrac{ax+b}{cx+d}}\right) dx$

An integral of the form

$$\int R\left(x, \sqrt[n]{\frac{ax+b}{cx+d}}\right) dx,$$

where $R(\cdot, \cdot)$ is a *rational function* of two variables, is rationalized via the substitution

$$t = \sqrt[n]{\frac{ax+b}{cx+d}}.$$

Example 8.1 (Rationalizing Substitution).

$$\int \frac{\sqrt{x+1}}{x} dx \qquad\qquad\qquad (a = b = d = 1,\ c = 0)$$

$$\text{substituting: } \boxed{t = \sqrt{x+1}, \; x = t^2 - 1, \; dx = 2t\,dt}\,;$$

$$= \int \frac{t}{t^2 - 1} 2t\,dt \qquad\qquad\qquad \text{by the \emph{integration rules};}$$

$$= 2 \int \frac{t^2}{t^2 - 1}\,dt \qquad \text{dividing: } \boxed{\frac{t^2}{t^2 - 1} = 1 + \frac{1}{t^2 - 1}} \;\; (\text{verify});$$

$$= 2 \int \left[1 + \frac{1}{t^2 - 1} \right] dt \qquad\qquad\qquad \text{by the \emph{integration rules};}$$

$$= 2 \left[\int 1\,dt + \int \frac{1}{t^2 - 1}\,dt \right]$$

we need not use the *Partial Fraction Method* for the second integral, it is a *"tall logarithm"* (see the *Table of Basic Integrals* (Appendix A));

$$= 2 \left[t + \frac{1}{2} \ln \left| \frac{t-1}{t+1} \right| \right] + C \qquad \text{\emph{substituting back} and simplifying;}$$

$$= 2\sqrt{x+1} + \ln \left| \frac{\sqrt{x+1} - 1}{\sqrt{x+1} + 1} \right| + C.$$

8.1.2.   Integrals of the Form $\displaystyle\int R\left(x, x^{m_1/n_1}, \ldots, x^{m_k/n_k}\right)\,dx$

An integral of the form

$$\int R\left(x, x^{m_1/n_1}, \ldots, x^{m_k/n_k}\right)\,dx,$$

where $R(\cdot, \ldots, \cdot)$ is a *rational function* of $k+1$ variables, is rationalized via the substitution

$$t = x^{1/n} \quad \text{or} \quad x = t^n,$$

where n is the *least common denominator* of all the fractions $\dfrac{m_1}{n_1}, \ldots, \dfrac{m_k}{n_k}$, which eliminates all the radicals.

Example 8.2 (Rationalizing Substitution).

$$\int \frac{1}{\sqrt[3]{x} + \sqrt[4]{x}}\,dx = \int \frac{1}{x^{1/3} + x^{1/4}}\,dx$$

$$\text{since 12 is the \emph{least common denominator} of } \frac{1}{3} \text{ and } \frac{1}{4},$$

$$\text{substituting: } \boxed{x = t^{12} \; (t \geq 0), \; dx = 12t^{11}\,dt}\,;$$

$$= \int \frac{12t^{11}}{[t^{12}]^{1/3} + [t^{12}]^{1/4}} \, dt \qquad \text{simplifying;}$$

$$= \int \frac{12t^{11}}{t^{12/3} + t^{12/4}} \, dt = \int \frac{12t^{11}}{t^4 + t^3} \, dt = \int \frac{12t^8}{t+1} \, dt$$

by the *integration rules* and *long/synthetic division* (*verify*);

$$= 12 \int \frac{t^8}{t+1} \, dt = 12 \int \left[t^7 - t^6 + t^5 - t^4 + t^3 - t^2 + t - 1 + \frac{1}{t+1} \right] dt$$

$$= 12 \left[\int t^7 \, dt - \int t^6 \, dt + \int t^5 \, dt - \int t^4 \, dt + \int t^3 \, dt - \int t^2 \, dt + \int t \, dt \right.$$

$$\left. - \int 1 \, dt + \int \frac{1}{t+1} \, dt \right]$$

$$= 12 \left[\frac{t^8}{8} - \frac{t^7}{7} + \frac{t^6}{6} - \frac{t^5}{5} + \frac{t^4}{4} - \frac{t^3}{3} + \frac{t^2}{2} - t + \ln|t+1| \right] + C$$

$$\text{substituting back:} \quad \boxed{t = x^{1/12}} \quad \text{and simplifying;}$$

$$= \frac{3}{2} x^{2/3} - \frac{12}{7} x^{7/12} + 2x^{1/2} - \frac{12}{5} x^{5/12} + 3x^{1/3} - 4x^{1/4} + 6x^{1/6}$$

$$- 12x^{1/12} + 12 \ln|x^{1/12} + 1| + C.$$

8.2. Integrals with Exponentials

An integral of the form

$$\int R(a^x) \, dx,$$

where $R(\cdot)$ is a *rational function*, $a > 0$, $a \neq 1$, is rationalized via the substitution

$$t = a^x \quad \text{or} \quad x = \log_a t.$$

Example 8.3 (Rationalizing Substitution).

$$\int \frac{1}{e^{3x} - e^x} \, dx \qquad (a = e) \quad \text{substituting:} \quad \boxed{t = e^x, \ x = \ln t, \ dx = \frac{1}{t} \, dt};$$

$$= \int \frac{1}{t^3 - t} \frac{1}{t} \, dt = \int \frac{1}{t^2(t^2 - 1)} \, dt$$

$$\text{transforming product into sum,}$$

$$\boxed{\frac{1}{t^2(t^2 - 1)} = \frac{1}{t^2 - 1} - \frac{1}{t^2}} \quad \text{(cf. Examples 2.10);}$$

$$= \int \left[\frac{1}{t^2-1} - \frac{1}{t^2} \right] dt \qquad \text{by the } integration\ rules;$$

$$= \int \frac{1}{t^2-1} - \int \frac{1}{t^2}\, dt$$

we need not use the *Partial Fraction Method* for the first integral, it is a *"tall logarithm"* (see the *Table of Basic Integrals* (Appendix A));

$$= \frac{1}{2} \ln \left| \frac{t-1}{t+1} \right| + \frac{1}{t} + C \qquad \text{substituting back;}$$

$$= \frac{1}{2} \ln \left| \frac{e^x-1}{e^x+1} \right| + e^{-x} + C.$$

8.3. Trigonometric Integrals

8.3.1. Integrals of the Form $\int R(\tan x)\, dx$

An integral of the form

$$\int R(\tan x)\, dx,$$

where $R(\cdot)$ is a *rational function*, is rationalized via the following substitution:

$$t = \tan x, \quad x = \arctan t, \quad dx = \frac{1}{t^2+1}\, dt.$$

Example 8.4 (Rationalizing Substitution).

$$\int \tan^5 x\, dx \qquad substituting: \quad \boxed{t = \tan x,\ x = \arctan t,\ dx = \frac{1}{t^2+1} dt}\,;$$

$$= \int \frac{t^5}{t^2+1}\, dt \quad \text{by } long\ division: \quad \boxed{\frac{t^5}{t^2+1} = t^3 - t + \frac{t}{t^2+1}} \quad (verify);$$

$$= \int \left[t^3 - t + \frac{t}{t^2+1} \right] dt \qquad \text{by the } integration\ rules;$$

$$= \int t^3\, dt - \int t\, dt + \int \frac{t}{t^2+1}\, dt = \frac{t^4}{4} - \frac{t^2}{2} + \int \frac{t}{t^2+1}\, dt$$

introducing the *missing constant* 2;

$$= \int t^3\, dt - \int t\, dt + \int \frac{t}{t^2+1}\, dt = \frac{t^4}{4} - \frac{t^2}{2} + \frac{1}{2} \int \frac{2t}{t^2+1}\, dt$$

$$\text{substituting:} \quad \boxed{u = t^2 + 1, \ du = 2tdt}$$

$$= \frac{t^4}{4} - \frac{t^2}{2} + \frac{1}{2} \int \frac{1}{u} \, du = \frac{t^4}{4} - \frac{t^2}{2} + \frac{1}{2} \ln|u| + C \qquad \text{substituting back;}$$

$$= \frac{\tan^4 x}{4} - \frac{\tan^2 x}{2} + \frac{1}{2} \ln(\tan^2 x + 1) + C$$

by the *trigonometric identity* $\tan^2 x + 1 = \sec^2 x$ and the *laws of logarithms*;

$$= \frac{\tan^4 x}{4} - \frac{\tan^2 x}{2} + \ln\sqrt{\sec^2 x} + C = \frac{\tan^4 x}{4} - \frac{\tan^2 x}{2} + \ln|\sec x| + C.$$

Remark 8.1. Applying *reduction formula* 5 of Appendix B *twice* yields the same result (*check*).

8.3.2. *Integrals of the Form* $\int R(\sin x, \cos x) \, dx$

An integral of the form

$$\int R(\sin x, \cos x) \, dx,$$

where $R(\cdot, \cdot)$ is a *rational function* of two variables, is rationalized via the following *universal substitution* (or *Weierstrass*[1] *substitution*):

$$t = \tan \frac{x}{2} \quad \text{or} \quad x = 2 \arctan t.$$

The substitution is *rationalizing* since

$$dx = \frac{2}{1 + t^2} dt,$$

and, by the *trigonometric identities* (Appendix C),

$$\sin x = \frac{2 \tan \frac{x}{2}}{1 + \tan^2 \frac{x}{2}} = \frac{2t}{1 + t^2}, \qquad \cos x = \frac{1 - \tan^2 \frac{x}{2}}{1 + \tan^2 \frac{x}{2}} = \frac{1 - t^2}{1 + t^2}.$$

Example 8.5 (Rationalizing Substitution).

$$\int \frac{1}{\cos x + 2} \, dx$$

$$\text{substituting:} \quad \boxed{\begin{array}{l} t = \tan \dfrac{x}{2}, \ x = 2 \arctan t, \ dx = \dfrac{2}{1 + t^2} dt \\ \cos x = \dfrac{1 - t^2}{1 + t^2} \end{array}} ;$$

---

[1] Karl Weierstrass (1815–1897)

$$= \int \frac{1}{\frac{1-t^2}{1+t^2}+2}\frac{2}{1+t^2}\,dt \quad \text{simplifying and using the \textit{integration rules};}$$

$$= \int \frac{2}{1-t^2+2(1+t^2)}\,dt = 2\int \frac{1}{t^2+(\sqrt{3})^2}\,dt = \frac{2}{\sqrt{3}}\arctan\frac{t}{\sqrt{3}}+C$$

$$\textit{substituting back;}$$

$$= \frac{2}{\sqrt{3}}\arctan\left(\frac{1}{\sqrt{3}}\tan\frac{x}{2}\right)+C.$$

Remark 8.2. Although the *universal substitution* always rationalizes the integrals of the above form, it may lead to cumbersome computations and is recommended to be avoided whenever possible. Thus,

- for the integral

$$\int \frac{\cos x}{\sin x + 1}\,dx,$$

we can avoid the *universal substitution* applying the simpler one: $\boxed{t = \sin x + 1}$;

- the integral

$$\int \frac{1}{1-\sin x}\,dx,$$

can be found *directly* (see Examples 5.1).

General Guidelines
on Avoiding the Universal Substitution

- If $R(-\sin x, \cos x) = -R(\sin x, \cos x)$ (*oddness* relative to $\sin x$), the substitution $\boxed{t = \cos x}$ can be used (cf. Sec. 5.2.2).
- If $R(\sin x, -\cos x) = -R(\sin x, \cos x)$ (*oddness* relative to $\cos x$), the substitution $\boxed{t = \sin x}$ can be used (cf. Sec. 5.2.2).
- If $R(-\sin x, -\cos x) = R(\sin x, \cos x)$ (*joint evenness* relative to $\sin x$ and $\cos x$), the substitution $\boxed{t = \tan x,\ x = \arctan t,\ dx = \frac{1}{t^2+1}\,dt}$ can be used considering that

$$\sin^2 x = \frac{\tan^2 x}{\tan^2 x + 1} = \frac{t^2}{t^2+1} \quad \text{and} \quad \cos^2 x = \frac{1}{\tan^2 x + 1} = \frac{1}{t^2+1}.$$

Example 8.6 (Avoiding the Universal Substitution).

$$\int \frac{\sin^2 x}{\sin^2 x + 1}\, dx \qquad \text{dividing;}$$

$$= \int \left[1 - \frac{1}{\sin^2 x + 1}\right] dx \qquad \text{by the } \textit{integration rules};$$

$$= \int 1\, dx - \int \frac{1}{\sin^2 x + 1}\, dx = x - \int \frac{1}{\sin^2 x + 1}\, dx$$

since the integrand is *jointly even* relative to $\sin x$ and $\cos x$,

substituting:
$$\boxed{t = \tan x,\ x = \arctan t,\ dx = \frac{1}{t^2 + 1} dt \\ \sin^2 x = \frac{t^2}{t^2 + 1}};$$

$$= x - \int \frac{1}{\frac{t^2}{t^2+1} + 1} \frac{1}{t^2+1}\, dt \qquad \text{simplifying and integrating;}$$

$$= x - \int \frac{1}{2t^2 + 1}\, dt = x - \int \frac{1}{(\sqrt{2}t)^2 + 1}\, dt = x - \frac{1}{\sqrt{2}} \arctan(\sqrt{2}t) + C$$

substituting back;

$$= x - \frac{1}{\sqrt{2}} \arctan(\sqrt{2}\tan x) + C.$$

8.4. Applications

Examples 8.7 (Applications).

1. Find the *area* of the region bounded by $y = \dfrac{1}{e^x + 1}$ and the x-axis on the interval $[0,1]$.

   Solution: Since $y = \dfrac{1}{e^x + 1} > 0$ on $[0,1]$,

$$A = \int_0^1 \frac{1}{e^x + 1}\, dx$$

substituting and *changing* the integration limits:

$$t = e^x, \quad x = \ln t, \quad dx = \frac{1}{t}\,dt \qquad \begin{array}{c|c} x & t \\ \hline 0 & 1 \\ 1 & e \end{array}\;;$$

$$= \int_1^e \frac{1}{(t+1)t}\,dt$$

transforming product into sum:

$$\frac{1}{(t+1)t} = \frac{1}{t} - \frac{1}{t+1} \qquad \text{(cf. Examples 2.10)};$$

$$= \int_1^e \left[\frac{1}{t} - \frac{1}{t+1}\right] dt \qquad\qquad \text{by the } \textit{integration rules};$$

$$= \int_1^e \frac{1}{t}\,dt - \int_1^e \frac{1}{t+1}\,dt \qquad\qquad \text{by the } \textit{Newton-Leibniz Formula};$$

$$= \ln|t| \Big|_1^e - \ln|t+1| \Big|_1^e = \ln e - \ln 1 - [\ln(e+1) - \ln 2]$$

considering that $\ln e = 1$ and $\ln 1 = 0$;

$$= 1 - \ln(e+1) + \ln 2 \text{ un.}^2.$$

2. Find the *volume* of the solid obtained by rotating the region bounded by $y = \dfrac{1}{\sqrt[3]{x+1}}$ and the *x-axis* on the interval $[0,7]$ about the *y-axis*.

Solution: Since the rotation axis is perpendicular to the axis of definition, by the *shell method*,

$$V = \int_0^7 2\pi x \frac{1}{\sqrt[3]{x+1}}\,dx \qquad\qquad \text{by the } \textit{integration rules};$$

$$= 2\pi \int_0^7 \frac{x}{\sqrt[3]{x+1}}\,dx$$

to rationalize, *substitute*:

$$t = \sqrt[3]{x+1}, \quad x = t^3 - 1, \quad dx = 3t^2\,dt \qquad \begin{array}{c|c} x & t \\ \hline 0 & 1 \\ 7 & 2 \end{array}\;;$$

$$= 2\pi \int_1^2 \frac{t^3 - 1}{t}\,3t^2\,dt \qquad\qquad \text{multiplying and dividing};$$

$$= 2\pi \int_1^2 3[t^4 - t]\,dt \qquad\qquad \text{by the } \textit{integration rules};$$

$$= 6\pi \int_1^2 [t^4 - t]\, dt = 6\pi \left[\int_1^2 t^4\, dt - \int_1^2 t\, dt \right]$$

by the *Newton-Leibniz Formula*;

$$= 6\pi \left[\frac{t^5}{5} \Big|_1^2 - \frac{t^2}{2} \Big|_1^2 \right] = 6\pi \left[\frac{1}{5}(2^5 - 1^5) - \frac{1}{2}(2^2 - 1^2) \right] = 6\pi \left[\frac{31}{5} - \frac{3}{2} \right]$$

$$= \frac{141\pi}{5} \text{ un.}^3.$$

Observe that the latter integral can be evaluated without being rationalized as follows:

$$V = \int_0^7 2\pi x \frac{1}{\sqrt[3]{x+1}}\, dx \qquad\qquad \text{by the } \textit{integration rules};$$

$$= 2\pi \int_0^7 \frac{x}{\sqrt[3]{x+1}}\, dx$$

substituting : $\quad t = x + 1,\ x = t - 1,\ dx = dt$

x	t
0	1
7	8

$$= 2\pi \int_1^8 \frac{t-1}{\sqrt[3]{t}}\, dt = 2\pi \int_1^8 \frac{t-1}{t^{1/3}}\, dt$$

dividing termwise and using the *laws of exponents* (see Appendix C);

$$= 2\pi \int_1^8 \left[\frac{t}{t^{1/3}} - \frac{1}{t^{1/3}} \right] dt = 2\pi \int_1^8 \left[t^{2/3} - t^{-1/3} \right] dt$$

by the *integration rules*;

$$= 2\pi \left[\int_1^8 t^{2/3}\, dt - \int_1^8 t^{-1/3}\, dt \right] \qquad \text{by the } \textit{Newton-Leibniz Formula};$$

$$= 2\pi \left[\frac{3}{5} t^{5/3} \Big|_1^8 - \frac{3}{2} t^{2/3} \Big|_1^8 \right] = 2\pi \left[\frac{3}{5}(8^{5/3} - 1^{5/3}) - \frac{3}{2}(8^{2/3} - 1^{2/3}) \right]$$

$$= 2\pi \left[\frac{93}{5} - \frac{9}{2} \right] = 2\pi \left[\frac{93}{5} - \frac{9}{2} \right] = \frac{141\pi}{5} \text{ un.}^3 \quad \text{(cf. Sec. 3.1.6)}.$$

8.5. Practice Problems

Evaluate the integrals.

1. $\displaystyle\int \frac{x+1}{\sqrt{2x+1}}\,dx$

2. $\displaystyle\int \frac{1}{\sqrt[3]{3x-1}+1}\,dx$

3. $\displaystyle\int \sqrt{\frac{x}{x+2}}\,dx$

4. $\displaystyle\int \frac{x^3}{\sqrt{x^2+6}}\,dx$

5. $\displaystyle\int \frac{\sqrt{x}}{\sqrt{x}+1}\,dx$

6. $\displaystyle\int \frac{1}{\sqrt{x}+\sqrt[3]{x}}\,dx$

7. $\displaystyle\int \frac{e^{3x}}{e^x-2}\,dx$

8. $\displaystyle\int \frac{e^{2x}+2e^x}{e^{2x}+1}\,dx$

9. $\displaystyle\int \tan^4 x\,dx$

10. $\displaystyle\int \frac{\sin^2 x}{\cos^4 x}\,dx$

11. $\displaystyle\int \frac{1}{1+3\cos^2 x}\,dx$

12. $\displaystyle\int \frac{1}{5+3\cos x}\,dx$

13. $\displaystyle\int \frac{1}{3\sin x+4\cos x}\,dx$

14. $\displaystyle\int \frac{1}{2\sin x-\cos x+5}\,dx$

Can We Integrate Them All Now?

Although the *Fundamental Theorem of Calculus (Part 1)* (Theorem 1.5) guarantees the existence of an *antiderivative F* for each function f *continuous* on an interval I, it says nothing about how to find it and whether it is feasible in terms of the known *elementary functions*.

Definition 8.2 (Elementary Function).
Elementary functions are

1. *Constants.*
2. *Powers* x^p.
3. *Exponentials* a^x and their *inverses*, i.e., *logarithms* $\log_a x$.
4. *Trigonometric functions* $\cos x$, $\sin x$, $\tan x$, $\cot x$, $\sec x$, $\csc x$ and their *inverses* $\arccos x$, $\arcsin x$, $\arctan x$, $\operatorname{arccot} x$, $\operatorname{arcsec} x$, $\operatorname{arccsc} x$.
5. *Hyperbolic functions* $\cosh x$, $\sinh x$, $\tanh x$, $\coth x$, $\operatorname{sech} x$, $\operatorname{csch} x$ and their inverses.
6. All the functions that can be obtained from the above via *combinations* (addition/subtraction, multiplication/division) and *compositions*.

Some *indefinite integrals*, although exist, cannot be evaluated in terms of the *elementary functions*, e.g.,

1. $\displaystyle\int \frac{e^x}{x}\, dx$ (*integral exponential*),

2. $\displaystyle\int \frac{1}{\ln x}\, dx$ (*integral logarithm*)

3. $\displaystyle\int \frac{\sin x}{x}\, dx$ (*integral sine*),

4. $\displaystyle\int \sin(e^x)\, dx$,

5. $\int e^{x^2}\,dx$ (*Poisson integral*),

6. $\int \sin(x^2)\,dx, \int \cos(x^2)\,dx$ (*Fresnel integrals*),

7. $\int \sqrt{x^3 + 1}\,dx.$

Remark 8.3. Observe that integral 2 can be reduced to integral 1 via the *substitution* $\boxed{x = e^t}$ and integral 4 to integral 3 via the *substitution* $\boxed{t = e^x}$ (*verify*).

Chapter 9

Improper Integrals

There are *two* essential restrictions in the theory of *definite integral* $\int_a^b f(x)\,dx$: both the *integration interval* and the *integrand* must be *bounded* (see Sec. 1.2.2).

The notion of *improper integral* is introduced in order to overcome these restrictions, i.e., to be able to integrate over *unbounded intervals* or have *unbounded integrands*. There are two types of improper integrals accordingly.

9.1. Type 1 Improper Integrals (Unbounded Interval)

Definition 9.1 (Type 1 Improper Integrals (Unbounded Interval)).
A *type 1 improper integral* is an integral over an *unbounded* interval.
Depending on the kind of the unboundedness of the integration interval (*right-sided, left-sided,* or *two-sided*), there are *three* cases:

$$\int_a^\infty f(x)\,dx, \quad \int_{-\infty}^b f(x)\,dx, \quad \text{or} \quad \int_{-\infty}^\infty f(x)\,dx,$$

where a and b are real numbers.

9.1.1. *Right-Sided Unboundedness*

Definition 9.2 (Improper Integral $\displaystyle\int_a^\infty f(x)\,dx$).
Suppose that a function $f : [a, \infty) \to \mathbb{R}$, where a is a real number, is *integrable* on the interval $[a, t]$, i.e., the *definite integral*

$$\int_a^t f(x)\,dx$$

exists (see Definition 1.6), for each $t \geq a$.

Remark 9.1. In particular, this is the case if f is *continuous* on $[a, \infty)$ (Corollary 1.1).

The *improper integral* of f from a to ∞

$$\int_a^\infty f(x)\,dx$$

is said to *converge* or to be *convergent* if the *limit*

$$\lim_{t\to\infty} \int_a^t f(x)\,dx$$

exists and is *finite*, in which case we say that the limit is the integral's *value* and write

$$\int_a^\infty f(x)\,dx = \lim_{t\to\infty} \int_a^t f(x)\,dx.$$

If the limit either *does not exists* or is *infinite*, the *improper integral* is said to *diverge* or to be *divergent*. In the latter case, we assign to the improper integral the *value* $\pm\infty$, respectively, and write

$$\int_a^\infty f(x)\,dx = \pm\infty.$$

Examples 9.1 (Improper Integrals).
Evaluate the *improper integral* or show that it *diverges*.

1. $\displaystyle\int_0^\infty \cos x\,dx$ is *divergent* since

$$\lim_{t\to\infty} \int_0^t \cos x\,dx = \lim_{t\to\infty} \sin x \Big|_0^t = \lim_{t\to\infty} [\sin t - \sin 0] = \lim_{t\to\infty} \sin t$$

does not exist.

2. $\displaystyle\int_0^\infty 1\,dx$ is *divergent* since

$$\lim_{t\to\infty} \int_0^t 1\,dx = \lim_{t\to\infty} x \Big|_0^t = \lim_{t\to\infty} [t - 0] = \lim_{t\to\infty} t = \infty.$$

Hence, $\displaystyle\int_0^\infty 1\,dx = \infty.$

3. $\int_1^\infty \dfrac{1}{\sqrt[3]{x}}\,dx$ is *divergent* since

$$\lim_{t\to\infty}\int_1^t x^{-1/3}\,dx = \lim_{t\to\infty}\frac{3}{2}x^{2/3}\Big|_1^t = \lim_{t\to\infty}\frac{3}{2}\left[t^{2/3}-1\right]=\infty.$$

Hence, $\int_1^\infty \dfrac{1}{\sqrt[3]{x}}\,dx = \infty.$

4. $\int_1^\infty \dfrac{1}{x^2}\,dx$ is *convergent* since

$$\lim_{t\to\infty}\int_1^t \frac{1}{x^2}\,dx = \lim_{t\to\infty}\left[-\frac{1}{x}\right]\Big|_1^t = \lim_{t\to\infty}\left[1-\frac{1}{t}\right]=1-0=1.$$

Hence, $\int_1^\infty \dfrac{1}{x^2}\,dx = 1.$

5. $\int_0^\infty x^2 e^{-x^3}\,dx$ is *convergent* since

$$\lim_{t\to\infty}\int_0^t x^2 e^{-x^3}\,dx \qquad\qquad \text{introducing the \emph{missing constant} 3;}$$

$$= \lim_{t\to\infty}\frac{1}{3}\int_0^t e^{-x^3}3x^2\,dx$$

substituting and *changing* the integration limits:

$$u=x^3,\ du=3x^2dx,\quad \begin{array}{c|c} x & u \\ \hline t & t^3 \\ 0 & 0 \end{array};$$

$$= \lim_{t\to\infty}\frac{1}{3}\int_0^{t^3} e^{-u}\,du \qquad\qquad \text{by the \emph{Newton-Leibniz Formula};}$$

$$= \lim_{t\to\infty}\frac{1}{3}\left[-e^{-u}\right]\Big|_0^{t^3} = \lim_{t\to\infty}\frac{1}{3}\left[1-e^{-t^3}\right]=\frac{1}{3}[1-0]=\frac{1}{3}.$$

Theorem 9.1 (Type 1 p-Integrals).
Let $a > 0$. Then the improper integral

$$\int_a^\infty \frac{1}{x^p}\,dx$$

converges for $p > 1$ and diverges for $p \leq 1$.

Proof.

(a) For $p > 1$,

$$\lim_{t \to \infty} \int_a^t x^{-p}\, dx = \lim_{t \to \infty} \frac{1}{-p+1} x^{-p+1} \Big|_a^t = \lim_{t \to \infty} \frac{1}{-p+1} \left[\frac{1}{t^{p-1}} - \frac{1}{a^{p-1}} \right]$$

considering that $p - 1 > 0$;

$$= \frac{1}{-p+1} \left[0 - \frac{1}{a^{p-1}} \right] = \frac{1}{(p-1)a^{p-1}}.$$

Hence, $\displaystyle \int_a^\infty \frac{1}{x^p}\, dx = \frac{1}{(p-1)a^{p-1}}.$

(b) For $p < 1$,

$$\lim_{t \to \infty} \int_a^t x^{-p}\, dx = \lim_{t \to \infty} \frac{1}{-p+1} x^{-p+1} \Big|_a^t = \lim_{t \to \infty} \frac{1}{-p+1} \left[t^{1-p} - a^{1-p} \right]$$

considering that $1 - p > 0$;

$$= \infty.$$

Hence, $\displaystyle \int_a^\infty \frac{1}{x^p}\, dx = \infty.$

(c) For $p = 1$,

$$\lim_{t \to \infty} \int_a^t \frac{1}{x}\, dx = \lim_{t \to \infty} \ln |x| \Big|_a^t = \lim_{t \to \infty} \left[\ln t - \ln a \right] = \infty.$$

Hence, $\displaystyle \int_a^\infty \frac{1}{x}\, dx = \infty.$

$\square$

9.1.2. *Left-Sided Unboundedness*

Definition 9.3 (Improper Integral $\displaystyle \int_{-\infty}^b f(x)\, dx$).
Suppose that a function $f : (-\infty, b] \to \mathbb{R}$, where b is a real number, is *integrable* on the interval $[t, b]$, i.e., the *definite integral*

$$\int_t^b f(x)\, dx$$

exists, for each $t \leq b$.

Remark 9.2. In particular, this is the case if f is *continuous* on $(-\infty, b]$.

The *improper integral* of f from $-\infty$ to b

$$\int_{-\infty}^{b} f(x)\,dx$$

is said to *converge* or to be *convergent* if the *limit*

$$\lim_{t \to -\infty} \int_{t}^{b} f(x)\,dx$$

exists and is *finite*, in which case we say that the limit is the integral's *value* and write

$$\int_{-\infty}^{b} f(x)\,dx = \lim_{t \to -\infty} \int_{t}^{b} f(x)\,dx.$$

If the limit either *does not exists* or is *infinite*, the *improper integral* is said to *diverge* or to be *divergent*. In the latter case, we assign to the improper integral the *value* $\pm\infty$, respectively, and write

$$\int_{-\infty}^{b} f(x)\,dx = \pm\infty.$$

Examples 9.2 (Improper Integrals).
Evaluate the *improper integral* or show that it *diverges*.

1. $\displaystyle\int_{-\infty}^{0} \sin x\,dx$ is *divergent* since

$$\lim_{t \to -\infty} \int_{t}^{0} \sin x\,dx = \lim_{t \to -\infty} \left[-\cos x\right]\Big|_{t}^{0} = \lim_{t \to -\infty} \left[\cos t - \cos 0\right]$$

$$= \lim_{t \to -\infty} \left[\cos t - 1\right] \quad \textit{does not exist.}$$

2. $\displaystyle\int_{-\infty}^{-1} \frac{1}{x^2}\,dx$ is *convergent* since

$$\lim_{t \to -\infty} \int_{t}^{-1} \frac{1}{x^2}\,dx = \lim_{t \to -\infty} \left[-\frac{1}{x}\right]\Big|_{t}^{-1} = \lim_{t \to \infty} \left[1 + \frac{1}{t}\right] = 1 + 0 = 1.$$

Hence, $\displaystyle\int_{-\infty}^{-1} \frac{1}{x^2}\,dx = 1.$

3. $\displaystyle\int_{-\infty}^{0} e^x \, dx$ is *convergent* since

$$\lim_{t\to-\infty} \int_{t}^{0} e^x \, dx = \lim_{t\to-\infty} e^x \Big|_{t}^{0} = \lim_{t\to-\infty} [1 - e^t] = 1 - 0 = 1.$$

Hence, $\displaystyle\int_{-\infty}^{0} e^x \, dx = 1.$

9.1.3.   Two-Sided Unboundedness

Definition 9.4 (Improper Integral $\displaystyle\int_{-\infty}^{\infty} f(x) \, dx$).
Suppose that a function $f : (-\infty, \infty) \to \mathbb{R}$ and, for any real $a < b$, is *integrable* on the interval $[a, b]$, i.e., the *definite integral*

$$\int_{a}^{b} f(x) \, dx$$

exists.

Remark 9.3. In particular, this is the case if f is *continuous* on $(-\infty, \infty)$.

If for some real a, both *improper integrals*

$$\int_{-\infty}^{a} f(x) \, dx \quad \text{and} \quad \int_{a}^{\infty} f(x) \, dx \tag{9.1}$$

converge, the *improper integral* of f from $-\infty$ to ∞ is said to *converge* or to be *convergent* and its *value* is defined as the sum of the values of the two one-sided improper integrals:

$$\int_{-\infty}^{\infty} f(x) \, dx = \int_{-\infty}^{a} f(x) \, dx + \int_{a}^{\infty} f(x) \, dx. \tag{9.2}$$

Remark 9.4. If both improper integrals in (9.1) converge for *some* real a, by the *additivity property* of *definite integral*, they converge for *any* real a and their sum in (9.2) is *independent* of the choice of a.

If for *some* real a, at least one of the one-sided improper integrals in (9.1) *diverges*, the two-sided improper integral

$$\int_{-\infty}^{\infty} f(x) \, dx$$

is said to *diverge* or to be *divergent*.

Examples 9.3 (Improper Integrals).
Evaluate the *improper integral* or show that it *diverges*.

1. $\displaystyle\int_{-\infty}^{\infty} \frac{1}{x^2+1}\, dx$ is *convergent* since

(a) $\displaystyle\int_{0}^{\infty} \frac{1}{x^2+1}\, dx = \lim_{t\to\infty}\int_{0}^{t} \frac{1}{x^2+1}\, dx = \lim_{t\to\infty} \arctan x\Big|_{0}^{t} = \lim_{t\to\infty}[\arctan t$

$- \arctan 0] = \dfrac{\pi}{2}$ *converges* and

(b) by the *symmetry* (the integrand is *even*),

$\displaystyle\int_{-\infty}^{0} \frac{1}{x^2+1}\, dx = \frac{\pi}{2}$ *converges* as well.

Hence, $\displaystyle\int_{-\infty}^{\infty} \frac{1}{x^2+1}\, dx = \int_{-\infty}^{0} \frac{1}{x^2+1}\, dx + \int_{0}^{\infty} \frac{1}{x^2+1}\, dx = \frac{\pi}{2}+\frac{\pi}{2} = \pi.$

2. $\displaystyle\int_{-\infty}^{\infty} e^x\, dx$ is *divergent* since

$\displaystyle\int_{0}^{\infty} e^x\, dx = \lim_{t\to\infty}\int_{0}^{t} e^x\, dx = \lim_{t\to\infty} e^x\Big|_{0}^{t} = \lim_{t\to\infty}[e^t - 1] = \infty,$

although (see Examples 9.2) $\displaystyle\int_{-\infty}^{0} e^x\, dx = 1.$

Hence, $\displaystyle\int_{-\infty}^{\infty} e^x\, dx = 1 + \infty = \infty.$

Geometric Interpretation

The value of a *type 1 improper integral*

$$\int_{a}^{\infty} f(x)\, dx, \quad \int_{-\infty}^{b} f(x)\, dx, \quad \text{or} \quad \int_{-\infty}^{\infty} f(x)\, dx,$$

when exists, *finite* or *infinite*, represents the *net area* of the region bounded by the graph of $f(x)$ and the *x-axis* on the corresponding interval.
When such a value does not exist, as, e.g., for the improper integral

$$\int_{0}^{\infty} \cos x\, dx,$$

the *net area* is *undefined*.

Theorem 9.2 (Comparison Test).
Suppose that functions $f, g : [a, \infty) \to \mathbb{R}$, where a is a real number, are

integrable on the interval $[a, t]$ for each $t \geq a$ and satisfy the inequality

$$0 \leq f(x) \leq cg(x)$$

with some $c > 0$ for all sufficiently large x-values.
Then

1. *If $\int_a^\infty g(x)\, dx$ is convergent, then so is $\int_a^\infty f(x)\, dx$.*

2. *If $\int_a^\infty f(x)\, dx$ is divergent, then so is $\int_a^\infty g(x)\, dx$.*

Remark 9.5. The analogous tests are in place for the cases of $(-\infty, b]$ and $(-\infty, \infty)$.

Example 9.4 (Using Comparison Test).
Determine whether the *improper integral*

$$\int_1^\infty \frac{1 + \sin x}{x \sqrt[10]{x}}\, dx$$

is *convergent* or *divergent*.

Solution: Since

$$-1 \leq \sin x \leq 1 \quad \text{for } x \geq 1,$$

then

$$0 \leq \frac{1 + \sin x}{x \sqrt[10]{x}} = \frac{1 + \sin x}{x^{1.1}} \leq \frac{2}{x^{1.1}} \quad \text{for } x \geq 1$$

and hence, by the *Comparison Test*, the *improper integral*

$$\int_1^\infty \frac{1 + \sin x}{x \sqrt[10]{x}}\, dx$$

converges along with the *improper integral*

$$\int_1^\infty \frac{1}{x^{1.1}}\, dx,$$

which is a *p-integral* with $p = 1.1 > 1$.

9.2. Type 2 Improper Integrals (Unbounded Integrand)

Definition 9.5 (Type 2 Improper Integrals (Unbounded Integrand)).
A *type 2 improper integral* is an integral of an *unbounded function* over a *bounded interval*.
Depending on the kind of the unboundedness of the integrand (at the *left endpoint*, at the *right endpoint*, or *inside the interval*), there are also *three* cases.

9.2.1. *Unboundedness at the Left Endpoint*

Definition 9.6 (Unboundedness at the Left Endpoint).
Suppose that a function $f : (a, b] \to \mathbb{R}$, where $a < b$ are real numbers, satisfy the following conditions

(a) $f(x) \to \infty$ or $f(x) \to -\infty$ as $x \to a+$;
(b) f is *integrable* on the interval $[t, b]$, i.e., the *definite integral*

$$\int_t^b f(x)\, dx$$

exists, for each $a < t \leq b$.

The *improper integral* of f from a to b

$$\int_a^b f(x)\, dx$$

is said to *converge* or to be *convergent* if the *limit*

$$\lim_{t \to a+} \int_t^b f(x)\, dx$$

exists and is *finite*, in which case we say that the limit is the integral's *value* and write

$$\int_a^b f(x)\, dx = \lim_{t \to a+} \int_t^b f(x)\, dx.$$

If the limit either *does not exists* or is *infinite*, the *improper integral* is said to *diverge* or to be *divergent*, in the latter case, we assign to the improper integral the *value* $\pm\infty$, respectively, and write

$$\int_a^b f(x)\, dx = \pm\infty.$$

Examples 9.5 (Improper Integrals).
Evaluate the *improper integral* or show that it *diverges*.

1. $\displaystyle\int_0^1 \frac{1}{\sqrt{x}}\, dx$ is *convergent* since

$$\lim_{t \to 0+} \int_t^1 x^{-1/2}\, dx = \lim_{t \to 0+} 2x^{1/2}\Big|_t^1 = \lim_{t \to 0+} 2[1 - t^{1/2}] = 2[1 - 0] = 2.$$

Hence, $\displaystyle\int_0^1 \frac{1}{\sqrt{x}}\, dx = 2.$

2. $\int_0^1 \dfrac{1}{x^2}\, dx$ is *divergent* since

$$\lim_{t \to 0+} \int_t^1 \frac{1}{x^2}\, dx = \lim_{t \to 0+} \left[-\frac{1}{x} \right]\Big|_t^1 = \lim_{t \to 0+} \left[\frac{1}{t} - 1 \right] = \infty.$$

Hence, $\displaystyle \int_0^1 \frac{1}{x^2}\, dx = \infty.$

Theorem 9.3 (Type 2 p-Integrals).
Let $b > 0$. Then the improper integral

$$\int_0^b \frac{1}{x^p}\, dx$$

converges for $p < 1$ and diverges for $p \geq 1$.

Proof.

(a) For $p < 1$,

$$\lim_{t \to 0+} \int_t^b x^{-p}\, dx = \lim_{t \to 0+} \frac{1}{-p+1} x^{-p+1}\Big|_t^b = \lim_{t \to 0+} \frac{1}{-p+1} \left[b^{1-p} - t^{1-p} \right]$$

considering that $1 - p > 0$;

$$= \frac{1}{-p+1}\left[b^{1-p} - 0 \right] = \frac{b^{1-p}}{1-p}.$$

Hence, $\displaystyle \int_0^b \frac{1}{x^p}\, dx = \frac{b^{1-p}}{1-p}.$

(b) For $p > 1$,

$$\lim_{t \to 0+} \int_t^b x^{-p}\, dx = \lim_{t \to 0+} \frac{1}{-p+1} x^{-p+1}\Big|_t^b = \lim_{t \to 0+} \frac{1}{-p+1} \left[\frac{1}{b^{p-1}} - \frac{1}{t^{p-1}} \right]$$

considering that $p - 1 > 0$;

$$= \infty.$$

Hence, $\displaystyle \int_0^b \frac{1}{x^p}\, dx = \infty.$

(c) For $p = 1$,

$$\lim_{t \to 0+} \int_t^b \frac{1}{x}\, dx = \lim_{t \to 0+} \ln |x| \Big|_t^b = \lim_{t \to 0+} [\ln b - \ln t] = \infty.$$

Hence, $\qquad \int_0^b \frac{1}{x}\, dx = \infty.$

$\square$

9.2.2. *Unboundedness at the Right Endpoint*

Definition 9.7 (Unboundedness at the Right Endpoint).

Suppose that a function $f : [a, b) \to \mathbb{R}$, where $a < b$ are real numbers, satisfies the following conditions

(a) $f(x) \to \infty$ or $f(x) \to -\infty$ as $x \to b-$;
(b) f is *integrable* on the interval $[a, t]$, i.e., the *definite integral*

$$\int_a^t f(x)\, dx$$

exists, for each $a \leq t < b$.

The *improper integral* of f from a to b

$$\int_a^b f(x)\, dx$$

is said to *converge* or to be *convergent* if the *limit*

$$\lim_{t \to b-} \int_a^t f(x)\, dx$$

exists and is *finite*, in which case we say that the limit is the integral's *value* and write

$$\int_a^b f(x)\, dx = \lim_{t \to b-} \int_a^t f(x)\, dx.$$

If the limit either *does not exists* or is *infinite*, the *improper integral* is said to *diverge* or to be *divergent*. In the latter case, we assign to the improper integral the *value* $\pm\infty$, respectively, and write

$$\int_a^b f(x)\, dx = \pm\infty.$$

Examples 9.6 (Improper Integrals).

Evaluate the *improper integral* or show that it *diverges*.

1. $\displaystyle\int_0^{\pi/2} \tan x \, dx$ *diverges* since

$$\int_0^{\pi/2} \tan x \, dx = \lim_{t\to\pi/2-} \int_0^t \tan x \, dx = \lim_{t\to\pi/2-} \ln|\sec x| \Big|_0^t$$

$$= \lim_{t\to\pi/2-} \ln|\sec t| \qquad\qquad \text{since} \quad \boxed{\sec t \to \infty \text{ as } t \to \pi/2-} ;$$

$$= \infty.$$

Hence, $\displaystyle\int_0^{\pi/2} \tan x \, dx = \infty.$

2. $\displaystyle\int_0^2 \frac{1}{\sqrt{2-x}} \, dx$ *converges* since

$$\lim_{t\to 2-} \int_0^t (2-x)^{-1/2} \, dx = \lim_{t\to 2-} \left[-2(2-x)^{1/2}\right]\Big|_0^t = \lim_{t\to 2-} 2[\sqrt{2} - \sqrt{2-t}]$$

$$= 2[\sqrt{2} - 0] = 2\sqrt{2}.$$

Hence, $\displaystyle\int_0^2 \frac{1}{\sqrt{2-x}} = 2\sqrt{2}.$

9.2.3. Unboundedness Inside the Interval

Definition 9.8 (Unboundedness Inside the Interval).

Suppose that a function f is defined on an interval $[a, b]$ $(-\infty < a < b < \infty)$, except, possibly, at a point $a < c < b$, and satisfies the following conditions:

(a) $f(x) \to \infty$ or $f(x) \to -\infty$ as $x \to c-$ and $x \to c+$;

(b) f is *integrable* on the interval $[a, t]$, i.e., the *definite integral*

$$\int_a^t f(x) \, dx$$

exists, for each $a \le t < c$.

(c) f is *integrable* on the interval $[t, b]$, i.e., the *definite integral*

$$\int_t^b f(x) \, dx$$

exists, for each $c < t \le b$.

If both improper integrals

$$\int_a^c f(x)\,dx \quad \text{and} \quad \int_c^b f(x)\,dx \qquad (9.3)$$

converge, the *improper integral* of f from a to b is said to *converge* or to be *convergent* and its *value* is defined as the sum

$$\int_a^b f(x)\,dx = \int_a^c f(x)\,dx + \int_c^b f(x)\,dx.$$

If at least one of the improper integrals in (9.3) *diverges*, the improper integral

$$\int_a^b f(x)\,dx$$

is said to *diverge* or to be *divergent*.

Examples 9.7 (Improper Integrals).
Evaluate the *improper integral* or show that it *diverges*.

1. $\int_{-1}^1 \dfrac{1}{x}\,dx$ has unboundedness at 0 and *diverges* since

$\int_0^1 \dfrac{1}{x}\,dx$ is a *divergent* type 2 p-integral with $p = 1 \geq 1$.

Thus, the *"solution"*:

$$\int_{-1}^1 \frac{1}{x}\,dx = \ln|x|\Big|_{-1}^1 = \ln 1 - \ln 1 = 0$$

is *incorrect*.

2. $\int_{-1}^2 \dfrac{1}{\sqrt[3]{x}}\,dx$ has unboundedness at 0 and *converges* since

(a) $\displaystyle\int_{-1}^0 \frac{1}{\sqrt[3]{x}}\,dx = \lim_{t\to 0-} \int_{-1}^t x^{-1/3}\,dx = \lim_{t\to 0-} \frac{3}{2}x^{2/3}\Big|_{-1}^t = \lim_{t\to 0-} \frac{3}{2}[t^{2/3} - 1]$

$= \dfrac{3}{2}[0-1] = -\dfrac{3}{2}$ is *convergent*,

(b) $\displaystyle\int_0^2 \frac{1}{\sqrt[3]{x}}\,dx = \lim_{t\to 0+} \int_t^2 x^{-1/3}\,dx = \lim_{t\to 0+} \frac{3}{2}x^{2/3}\Big|_t^2 = \lim_{t\to 0+} \frac{3}{2}[2^{2/3} - t^{2/3}]$

$= \dfrac{3}{2}[2^{2/3} - 0] = \dfrac{3}{2}2^{2/3}$ is *convergent*.

Hence, $\displaystyle\int_{-1}^{2}\frac{1}{\sqrt[3]{x}}\,dx = \int_{-1}^{0}\frac{1}{\sqrt[3]{x}}\,dx + \int_{0}^{2}\frac{1}{\sqrt[3]{x}}\,dx = -\frac{3}{2}+\frac{3}{2}2^{2/3} = \frac{3}{2}[2^{2/3}-1].$

Remark on Using Newton-Leibniz Formula

As the example of the integral

$$\int_{-1}^{1}\frac{1}{x}\,dx$$

shows, in the case of a type 2 improper integral with unboundedness inside the interval, the *Newton-Leibniz Formula* (Theorem 1.6) *cannot* be used without first checking whether the integral converges. If it does, the formula works just like for a proper definite integral.

Thus, in the prior example,

$$\int_{-1}^{2}\frac{1}{\sqrt[3]{x}}\,dx = \int_{-1}^{2} x^{-1/3}\,dx = \frac{3}{2}x^{2/3}\Big|_{-1}^{2} = \frac{3}{2}\left[2^{2/3}-1\right].$$

Geometric Interpretation

The value of a *type 2 improper integral*

$$\int_{a}^{b} f(x)\,dx,$$

when exists, *finite* or *infinite*, represents the *net area* of the region bounded by the graph of $f(x)$ and the x-axis on the interval $[a, b]$.
When such a value does not exist, the *net area* is *undefined*.

If $f(x) \geq 0$ on the interval, the value of the *type 2 improper integral* when exists, *finite* or *infinite*, represents the *area* of the region bounded by the graph of $f(x)$ and the x-axis on the corresponding interval.

Remark 9.6. The analogue of the *Comparison Test* for *type 1* improper integral also holds for *type 2* improper integral.

9.3. Applications

Examples 9.8 (Applications).

1. Find the *volume* of *"Gabriel's horn"*, i.e., the solid obtained by rotating the region bounded by $y = \dfrac{1}{x}$ and the x-axis on the interval $[1, \infty)$ about the x-axis.

   Solution: By the *disk method*,

   $$A = \int_1^\infty \pi \left[\frac{1}{x}\right]^2 dx = \int_1^\infty \pi \frac{1}{x^2} dx \qquad \text{by the } \textit{integration rules};$$

   $$= \pi \int_1^\infty \frac{1}{x^2} dx \qquad \text{since } \int_1^\infty \frac{1}{x^2} dx = 1 \quad \text{(see Examples 9.1)};$$

   $$= \pi \text{ un.}^2.$$

2. Find the *surface area* of the solid of revolution of the preceding problem.

   Solution: By the *area of the surface of revolution formula*,

   $$A = \int_1^\infty 2\pi \frac{1}{x} \sqrt{1 + \left[\left(\frac{1}{x}\right)'\right]^2} dx = \int_1^\infty 2\pi \frac{1}{x} \sqrt{1 + \left[-\frac{1}{x^2}\right]^2} dx$$

   $$\text{simplifying and using the } \textit{integration rules};$$

   $$= 2\pi \int_1^\infty \frac{\sqrt{x^4 + 1}}{x^3} dx = \infty.$$

   Indeed, since $\dfrac{\sqrt{x^4+1}}{x^3} \geq \dfrac{\sqrt{x^4}}{x^3} = \dfrac{x^2}{x^3} = \dfrac{1}{x}$ on $[1, \infty)$,

   $$\int_1^\infty \frac{\sqrt{x^4+1}}{x^3} dx \geq \int_1^\infty \frac{1}{x} dx = \infty,$$

   the latter being a *divergent* type 1 p-integral with $p = 1$.

   Remark 9.7. The two latter examples demonstrate a somewhat counterintuitive fact that the surface area of a solid with a finite volume may be infinite.

3. Find the *volume* of the solid generated by revolving the region bounded by $y = e^{-x}$ and the *x-axis* on the interval $[0, \infty)$ about the *y-axis*.

   Solution: Since the rotation axis is perpendicular to the axis of definition, by the *shell method*,

$$V = \int_0^\infty 2\pi x e^{-x}\, dx = \lim_{t\to\infty} \int_0^t 2\pi x e^{-x}\, dx \qquad \text{by the } \textit{integration rules};$$

$$= \lim_{t\to\infty} 2\pi \int_0^t x e^{-x}\, dx$$

by the *type 1* partial integration scenario,

$$\boxed{\begin{array}{ll} u = x, & du = dx \\ dv = e^{-x}dx, & v = \int e^{-x}\, dx = -e^{-x} \end{array}};$$

$$= \lim_{t\to\infty} 2\pi \left[-xe^{-x} \Big|_0^t + \int_0^t e^{-x}\, dx \right] = \lim_{t\to\infty} 2\pi \left[-te^{-t} - e^{-x} \Big|_0^t \right]$$

$$= \lim_{t\to\infty} 2\pi \left[-te^{-t} - e^{-t} + 1 \right] \qquad \text{by the } \textit{limit laws};$$

$$= 2\pi \left[-\lim_{t\to\infty} te^{-t} - \lim_{t\to\infty} e^{-t} + 1 \right] = 2\pi\left[-0 - 0 + 1 \right] = 2\pi \text{ un.}^3.$$

Indeed,

$$\lim_{t\to\infty} e^{-t} = 0$$

and

$$\lim_{t\to\infty} te^{-t} = \{\infty \cdot 0\} = \lim_{t\to\infty} \frac{t}{e^t} = \left\{\frac{\infty}{\infty}\right\}$$

by *L'Hôpital's Rule* (see, e.g., [1, 6]);

$$= \lim_{t\to\infty} \frac{t'}{[e^t]'} = \lim_{t\to\infty} \frac{1}{e^t} = \frac{1}{\infty} = 0.$$

9.4.   Practice Problems

Evaluate the *improper integral* or show that it *diverges*.

1. $\displaystyle \int_2^\infty \frac{1}{x^2 + x - 2}\, dx$

2. $\displaystyle \int_e^\infty \frac{1}{x \ln x}\, dx$

3. $\displaystyle \int_0^\infty \frac{x}{\sqrt{x^2 + 1}}\, dx$

4. $\displaystyle \int_0^\infty e^{-x} \cos x\, dx$

5. $\displaystyle \int_{-\infty}^{-1} \frac{\sin(\pi/x)}{x^2}\, dx$

6. $\displaystyle \int_{-\infty}^\infty \frac{1}{(x^2 + x + 1)^2}\, dx$

7. $\int_1^2 \frac{1}{\sqrt{x-1}}\, dx$

8. $\int_0^1 \frac{x^2}{x^3-1}\, dx$

9. $\int_{-3}^3 \frac{1}{\sqrt{9-x^2}}\, dx$

10. $\int_1^3 \frac{1}{(x-2)^{2/3}}\, dx$

11. $\int_{-\pi/2}^0 \sec x\, dx$

12. $\int_0^1 \ln x\, dx$

Mixed Integration Problems

Evaluate the integrals.

1. $\displaystyle\int \frac{1}{(x+1)\sqrt{x^2+1}}\,dx$

2. $\displaystyle\int \frac{x^2-x+1}{\sqrt[3]{x}}\,dx$

3. $\displaystyle\int (1-2x)^9\,dx$

4. $\displaystyle\int \frac{1}{1+\tan x}\,dx$

5. $\displaystyle\int \sqrt{\frac{1+x}{1-x}}\,dx$

6. $\displaystyle\int x\ln(x^4+4)\,dx$

7. $\displaystyle\int \sin x \cos(\cos x)\,dx$

8. $\displaystyle\int \frac{\sin(\ln x)}{x}\,dx$

9. $\displaystyle\int \frac{e^{-\sqrt{x}}}{\sqrt{x}}\,dx$

10. $\displaystyle\int e^{2x}\cos 3x\,dx$

11. $\displaystyle\int x\arctan x\,dx$

12. $\displaystyle\int \frac{e^x}{1-3e^x}\,dx$

13. $\displaystyle\int \frac{x^3}{\sqrt{x^2+1}}\,dx$

14. $\displaystyle\int \sin(\ln x)\,dx$

15. $\displaystyle\int \tan^3 x\,dx$
(without *Reduction Formulas*)

16. $\displaystyle\int \sin 3x \sin 5x\,dx$

17. $\displaystyle\int x\cos^2 x\,dx$

18. $\displaystyle\int x\cot^2 x\,dx$

19. $\displaystyle\int \cos^4 x\,dx$
(without *Reduction Formulas*)

20. $\displaystyle\int \frac{1}{(\sin x+\cos x)^2}\,dx$

21. $\displaystyle\int \frac{1}{x^4+4}\,dx$

22. $\displaystyle\int \frac{x}{x^4+3x^2+2}\,dx$

23. $\displaystyle\int \frac{\cos^2 x}{\sin x}\,dx$

24. $\displaystyle\int \frac{\sin^2 x}{\cos^4 x}\,dx$

25. $\displaystyle\int \frac{1}{(1+\sqrt{x})^3}\,dx$

26. $\displaystyle\int \frac{1}{\sqrt{x(2-x)}}\,dx$

27. $\displaystyle\int \frac{1}{e^{3x} - e^x}\,dx$

28. $\displaystyle\int \frac{1}{\sqrt{e^x + 1}}\,dx$

29. $\displaystyle\int \frac{\sqrt{x^2 + 2x}}{x^3}\,dx$

30. $\displaystyle\int \frac{1}{x\sqrt{x^3 - 1}}\,dx$

Answer Key

Chapter 2: Direct Integration

1. $\frac{2}{5}\sqrt{5x-6}+C$

2. $-\frac{1}{33}(2-3x)^{11}+C$

3. $x^3+x^2-\ln|x|+C$

4. $e^x+\frac{1}{x}+C$

5. $-\frac{1}{x}-\frac{2}{3x^3}+C$

6. $x^2+3x+4\ln|x-1|+C$

7. $\frac{2}{3}x^{3/2}+6\sqrt{x}-3x-\ln|x|+C$

8. $\frac{3}{112}(2x-1)^{4/3}(8x+3)+C$

9. $\frac{3}{2}x-\frac{5}{4}\ln|2x+3|+C$

10. $\frac{1}{3}x^3-2x+2\sqrt{2}\arctan\frac{x}{\sqrt{2}}+C$

11. $x^3-3x^2+7x-12\ln|x+2|+C$

12. $\frac{4^x}{\ln 4}+2\frac{6^x}{\ln 6}+\frac{9^x}{\ln 9}+C$

13. $\ln|x|-\frac{1}{4x^4}+C$

14. $\arcsin x+\ln(x+\sqrt{x^2+1})+C$
    $=\arcsin x+\operatorname{arcsinh}x+C$

15. $\frac{1}{6}(x+4)^{3/2}-\frac{1}{6}x^{3/2}+C$

16. $\frac{1}{4}\ln\left|\frac{x-1}{x+3}\right|+C$

17. $\frac{1}{2\sqrt{2}}\arctan\frac{1+2x}{\sqrt{2}}+C$

18. $\arcsin\frac{x+2}{3}+C$

19. $\frac{1}{\sqrt{3}}\ln\left|3x-1+\sqrt{9x^2-6x-3}\right|$
    $+C$

20. $\tan(x)-\sin(x)+C$

21. $-\cot x-\tan x=-2\csc 2x+C$

22. $x-\frac{1}{2}\cos 2x+C$

23. $\frac{1}{2}x-\frac{1}{2}\sin(x+2)+C$

24. $\frac{1}{4}\sin 2x+\frac{1}{16}\sin 8x+C$

25. $-\cot x+\csc x+C=\tan\frac{x}{2}+C$

26. $\operatorname{sgn}(\cos x+\sin x)(\sin x-\cos x)$
    $+C$

27. $\frac{1}{7}$

28. $\ln 4-2$

29. $\frac{5}{6}$

30. $1-\frac{\pi}{4}$

Chapter 3: Method of Substitution

1. $2\sqrt{x^2 - 3x + 4} + C$

2. $\ln\left|e^x + \sqrt{e^{2x} - 1}\right| + C$

3. $\dfrac{1}{2}\arcsin^2 x + C$

4. $\ln|\ln x - 3| + C$

5. $\ln|\ln|\ln x|| + C$

6. $\dfrac{1}{2}\ln(x^2 + 2) + C$

7. $\dfrac{3}{2}\ln|x^2 - 5| + \dfrac{1}{\sqrt{5}}\ln\left|\dfrac{x - \sqrt{5}}{x + \sqrt{5}}\right|$ $+ C$

8. $-\dfrac{\cos 2x}{4} + C$

9. $\dfrac{1}{2}\tan^2 x + C$ or $\dfrac{1}{2}\sec^2 x + C$

10. $\dfrac{3}{2}(\sin x - \cos x)^{2/3} + C$

11. $-\dfrac{1}{\sqrt{2}}\ln\left|\sqrt{2}\cos x + \sqrt{\cos 2x}\right| + C$

12. $\dfrac{1}{3}e^{x^3} + C$

13. $\dfrac{2}{\ln 5}5^{\sqrt{x}} + C$

14. $-\dfrac{1}{16}(1 - 2x^6)^{4/3} + C$

15. $\dfrac{1}{216}(1 - 3x^2)^{12}$ $- \dfrac{1}{198}(1 - 3x^2)^{11} + C$

16. $\dfrac{1}{6}\arctan\dfrac{x^3}{2} + C$

17. $\dfrac{1}{4}\ln\left|x^4 + \sqrt{x^8 - 1}\right| + C$

18. $\dfrac{1}{2}\ln(x^2 + x + 1) - \dfrac{1}{\sqrt{3}}\arctan\dfrac{2x + 1}{\sqrt{3}}$ $+ C$

19. $-\sqrt{1 - x^2} - \arcsin x + C$

20. $-(1 - x^2)^{1/2} + \dfrac{2}{3}(1 - x^2)^{3/2}$ $- \dfrac{1}{5}(1 - x^2)^{5/2} + C$

21. $\dfrac{1}{2}\ln\left|\dfrac{\sqrt{x^2 + 1} - 1}{\sqrt{x^2 + 1} + 1}\right| + C$

22. $\dfrac{1}{\sqrt{2}}\arctan\dfrac{\tan x}{\sqrt{2}} + C$

23. $\cos(1/x) + C$

24. $\ln\left|\dfrac{\sqrt{e^x + 1} - 1}{\sqrt{e^x + 1} + 1}\right| + C$

25. $\dfrac{\pi^2}{32}$

26. $\ln(4/3)$

Chapter 4: Method of Integration by Parts

1. $-x^3 e^{-x} - 3x^2 e^{-x} - 6xe^{-x}$ $- 6e^{-x} + C$

2. $x^2 \sin x + 2x \cos x - 2\sin x + C$

3. $x \ln x - x + C$

4. $\dfrac{x^2}{2}\operatorname{arcsec} x - \operatorname{sgn} x\dfrac{\sqrt{x^2 - 1}}{2} + C$

5. $\dfrac{2}{13}e^{2x}\sin 3x - \dfrac{3}{13}e^{2x}\cos 3x + C$

6. $-\dfrac{1}{x}\ln x - \dfrac{1}{x} + C$

7. $-\dfrac{1}{4}x\cos 2x + \dfrac{1}{8}\sin 2x + C$

8. $x \arctan x - \dfrac{1}{2}\ln(1 + x^2) + C$

9. $\dfrac{1}{2}x\cos(\ln x) + \dfrac{1}{2}x\sin(\ln x) + C$

10. $x(\ln x)^2 - 2x \ln x + 2x + C$

11. $-x\cot x + \ln|\sin x| + C$

12. $-\dfrac{1}{2}x^2 e^{-x^2} - \dfrac{1}{2}e^{-x^2} + C$

13. $\dfrac{x^2}{2}(\arctan x)^2 - x \arctan x$
$+ \dfrac{1}{2}(\arctan x)^2 + \dfrac{1}{2}\ln(1+x^2) + C$

14. $-\dfrac{1}{x}\arcsin x - \ln\left|\dfrac{1+\sqrt{1-x^2}}{x}\right|$
$+ C$

15. $3\ln 3 - 2$

16. $-1/2$

17. $\dfrac{1}{2}e^\pi + \dfrac{1}{2}$

18. $\dfrac{8}{9}e^3 + \dfrac{4}{9}$

19. $\sqrt{3}/2 + \pi/12$

20. $\pi/9 + \sqrt{3}/3$

Chapter 5: Trigonometric Integrals

1. $-\dfrac{1}{14}\cos 7x - \dfrac{1}{6}\cos 3x + C$

2. $\dfrac{1}{2}x + \dfrac{1}{12}\sin 6x + C$

3. $3x - 4\cos x - \sin 2x + C$

4. $\dfrac{3}{2}x - \sin 2x + \dfrac{1}{8}\sin 4x + C$

5. $\dfrac{x}{8} - \dfrac{\sin 4x}{32} + C$

6. $\dfrac{1}{16}x - \dfrac{1}{64}\sin 4x + \dfrac{1}{48}\sin^3 2x + C$

7. $-\dfrac{\cos^3 x}{3} + \dfrac{\cos^5 x}{5} + C$

8. $\dfrac{\sin^3 x}{3} - \dfrac{\sin^5 x}{5} + C$

9. $-\dfrac{1}{7}\sin^6 x \cos x - \dfrac{6}{35}\sin^4 x \cos x$
$- \dfrac{8}{35}\sin^2 x \cos x - \dfrac{16}{35}\cos x + C$

10. $\dfrac{1}{4}\cos^3 x \sin x + \dfrac{3}{8}\cos x \sin x$

$+ \dfrac{3}{8}x + C$

11. $\sec x + \cos x + C$

12. $3\sin^{1/3} x - \dfrac{3}{7}\sin^{7/3} x + C$

13. $-\dfrac{1}{2\tan^2 x} + \ln|\tan x| + C$

14. $-\dfrac{\csc^7 x}{7} + \dfrac{\csc^5 x}{5} + C$

15. $\dfrac{1}{4}\sec^4 x - \tan^2 x - \ln|\cos x| + C$

16. $\dfrac{1}{2}\ln|\sec(\ln x) + \tan(\ln x)|$
$+ \dfrac{1}{2}\sec(\ln x)\tan(\ln x) + C$

17. $\dfrac{8}{9\sqrt{3}} + \dfrac{\pi}{6}$

18. $1/3$

19. $\sqrt{2}$

20. $\dfrac{1\cdot 3\cdot 5\cdots(2n-1)}{2\cdot 4\cdot 6\cdots 2n}\pi$

Chapter 6: Trigonometric Substitutions

1. $\sqrt{3}/12$

2. $\dfrac{\pi}{72} + \dfrac{2-\sqrt{3}}{24}$

3. $\pi/12$

4. $\dfrac{\sqrt{2}}{2} + \dfrac{1}{2}\ln(\sqrt{2}+1)$

5. $-\dfrac{1}{15}(3x^4 + 4x^2 + 8)\sqrt{1-x^2}$
$+ C$

6. $\dfrac{1}{2}x\sqrt{x^2+2} - \ln\left|x+\sqrt{x^2+2}\right|$
$+ C$

7. $\dfrac{1}{2}x\sqrt{x^2-9} + \dfrac{9}{2}\ln\left|x+\sqrt{x^2-9}\right|$
$+ C$

8. $-\dfrac{\sqrt{9-4x^2}}{x} - 2\arcsin\dfrac{2}{3}x + C$

9. $\dfrac{9}{8}\operatorname{sgn} x \operatorname{arcsec}\dfrac{3x}{4} - \dfrac{\sqrt{9x^2-16}}{2x^2}$
$+ C$

10. $-\arctan\dfrac{1}{\sqrt{x^2-2}}$
$+ \arctan\dfrac{\sqrt{x^2-2}}{x} + C$

11. $\dfrac{1}{2}\sqrt{x^2+2x+10}\,(x+1)$
$- \dfrac{9}{2}\ln\left|\sqrt{x^2+2x+10}+x+1\right|$
$+ C$

12. $\dfrac{\sqrt{x^2+2x-5}}{6(x+1)} + C$

13. $\dfrac{1}{2}\ln\dfrac{x^2(\sqrt{x^4+1}+x^2)}{\sqrt{x^4+1}+1} + C$

15. $\dfrac{9}{8}\arcsin\dfrac{2x-1}{3}$
$+ \dfrac{1}{4}(2x-1)\sqrt{2+x-x^2} + C$

16. $\dfrac{1}{4}(x^2+1)\sqrt{x^4+2x^2-1}$
$- \dfrac{1}{2}\ln(x^2+1+\sqrt{x^4+2x^2-1})$
$+ C$

Chapter 7: Integration of Rational Functions

1. $\dfrac{1}{7}\ln|x+2| + \dfrac{13}{7}\ln|x-5| + C$

2. $\dfrac{2}{3}\ln|x+2| + \dfrac{1}{3}\ln|x-1| + C$

3. $\dfrac{1}{6}\ln\dfrac{x^2+1}{x^2+4} + C$

4. $\dfrac{1}{3}\ln|x+1| - \dfrac{1}{6}\ln|x^2-x+1|$
$+ \dfrac{1}{\sqrt{3}}\arctan\dfrac{2x-1}{\sqrt{3}} + C$

5. $x + \dfrac{1}{6}\ln|x| - \dfrac{9}{2}\ln|x-2|$
$+ \dfrac{28}{3}\ln|x-3| + C$

6. $4\ln|x+2| - \dfrac{4}{x+2} - 4\ln|x+1|$
$- \dfrac{1}{x+1} + C$

7. $\dfrac{3}{8}\arctan x + \dfrac{x}{8(x^2+1)}$

$+ \dfrac{3}{16}\ln\left|\dfrac{x+1}{x-1}\right| - \dfrac{1/16}{x-1} - \dfrac{1/16}{x+1}$
$+ C$

8. $x - 2\ln|x+1| - \dfrac{1}{x+1} + C$

9. $-\dfrac{1}{96(x-1)^{96}} - \dfrac{3}{97(x-1)^{97}}$
$- \dfrac{3}{98(x-1)^{98}} - \dfrac{1}{99(x-1)^{99}} + C$

10. $\dfrac{1}{4}x^4 + \dfrac{1}{4}\ln(x^4+1) - \ln(x^4+2) + C$

11. $\dfrac{1}{10}\left[\ln x^{10} - \ln|x^{10}-1| - \dfrac{1}{x^{10}-1}\right]$
$+ C$

12. $\ln|x| - \dfrac{2}{7}\ln|x^7+1| + C$

13. $\dfrac{1}{5}\ln\left|\dfrac{x^5+5x}{x^5+5x+1}\right| + C$

14. $\dfrac{1}{\sqrt{5}}\arctan\dfrac{x^2-1}{\sqrt{5}x} + C$

Chapter 8: Rationalizing Substitutions

1. $\frac{1}{6}(2x+1)^{3/2} + \frac{1}{2}(2x+1)^{1/2} + C$

2. $\frac{1}{2}(3x-1)^{2/3} - (3x-1)^{1/3}$
$+ \ln\left|(3x-1)^{1/3}+1\right| + C$

3. $\ln\left|\sqrt{\dfrac{x}{x+2}} - 1\right|$
$- \ln\left|\sqrt{\dfrac{x}{x+2}} + 1\right|$
$+ (x+2)\sqrt{\dfrac{x}{x+2}} + C$

4. $\frac{1}{3}(x^2+6)^{3/2} - 6(x^2+6)^{1/2} + C$

5. $x - 2\sqrt{x} + 2\ln(\sqrt{x}+1) + C$

6. $6x^{1/6} - 3x^{1/3} + 2x^{1/2} - 6\ln(x^{1/6} + 1) + C$

7. $\frac{1}{2}e^{2x} + 2e^x + 4\ln|e^x - 2| + C$

8. $\frac{1}{2}\ln(e^{2x}+1) + 2\arctan e^x + C$

9. $\frac{1}{3}\tan^3 x - \tan x + x + C$

10. $\frac{1}{3}\tan^3 x + C$

11. $\frac{1}{2}\arctan\left(\tan\dfrac{x}{2} - \dfrac{1}{\tan\dfrac{x}{2}}\right) + C$

12. $\frac{1}{2}\arctan\left(\dfrac{1}{2}\tan\dfrac{x}{2}\right) + C$

13. $\frac{2}{5}\ln\left|\dfrac{1 + 2\tan\dfrac{x}{2}}{\sqrt{4 + 6\tan\dfrac{x}{2} - 4\tan^2\dfrac{x}{2}}}\right| + C$

14. $\frac{1}{\sqrt{5}}\arctan\left(\dfrac{3}{\sqrt{5}}\tan\dfrac{x}{2} + \dfrac{1}{\sqrt{5}}\right) + C$

Chapter 9: Improper Integrals

1. $\frac{2}{3}\ln 2$

2. The integral *diverges.*

3. The integral *diverges.*

4. 1/2

5. $-2/\pi$

6. $\dfrac{4\sqrt{3}}{9}\pi$

7. 2

8. The integral *diverges.*

9. π

10. 6

11. The integral *diverges.*

12. -1

Mixed Integration Problems

1. $-\dfrac{1}{\sqrt{2}}\ln\left|\dfrac{\sqrt{2}\sqrt{x^2+1} - x + 1}{x+1}\right| + C$
$= \dfrac{1}{\sqrt{2}}\ln\left|\dfrac{\sqrt{2}\sqrt{x^2+1} + x - 1}{x+1}\right| + C$

2. $\dfrac{3}{8}x^{8/3} - \dfrac{3}{5}x^{5/3} + \dfrac{3}{2}x^{2/3} + C$

3. $-\dfrac{1}{20}(1-2x)^{10} + C$

4. $\frac{1}{2}\ln|\tan x + 1|$

$$-\frac{1}{4}\ln\left(\tan^2 x + 1\right) + \frac{1}{2}x + C$$

$$= \frac{1}{2}x + \frac{1}{2}\ln|\cos x + \sin x| + C$$

5. $2\arctan\left(\sqrt{\dfrac{x+1}{1-x}}\right)$

$\quad - (1-x)\sqrt{\dfrac{x+1}{1-x}} + C$

6. $\dfrac{x^2}{2}\ln(x^4 + 4) - x^2 + 2\arctan\dfrac{x^2}{2}$
$\quad + C$

7. $-\sin\left(\cos x\right) + C$

8. $-\cos(\ln x) + C$

9. $-2e^{-\sqrt{x}} + C$

10. $\dfrac{e^{2x}}{13}\left[2\cos 3x + 3\sin 3x\right] + C$

11. $\dfrac{x^2}{2}\arctan x - \dfrac{1}{2}x + \dfrac{1}{2}\arctan x + C$

12. $-\dfrac{1}{3}\ln|1 - 3e^x| + C$

13. $\dfrac{(x^2+1)^{3/2}}{3} - \sqrt{x^2+1} + C$

14. $\dfrac{x}{2}\left[\sin(\ln x) - \cos(\ln x)\right] + C$

15. $\dfrac{\tan^2 x}{2} - \ln|\sec x| + C$

16. $\dfrac{1}{4}\sin 2x - \dfrac{1}{16}\sin 8x + C$

17. $\dfrac{1}{4}x^2 + \dfrac{1}{4}x\sin 2x + \dfrac{1}{8}\cos 2x + C$

18. $-\dfrac{x^2}{2} - x\cot x + \ln|\sin x| + C$

19. $\dfrac{3x}{8} + \dfrac{\sin 2x}{4} + \dfrac{\sin 4x}{32} + C$

20. $-\dfrac{1}{\tan x + 1} + C$

21. $-\dfrac{1}{16}\ln(x^2 - 2x + 2)$
$\quad + \dfrac{1}{8}\arctan(x - 1)$
$\quad + \dfrac{1}{16}\ln(x^2 + 2x + 2)$
$\quad + \dfrac{1}{8}\arctan(x + 1) + C$

22. $\dfrac{1}{2}\left[\ln(x^2 + 1) - \ln(x^2 + 2)\right] + C$

23. $\ln|\csc x - \cot x| + \cos x + C$

24. $\dfrac{\tan^3 x}{3} + C$

25. $-\dfrac{2}{1 + \sqrt{x}} + \dfrac{1}{(1 + \sqrt{x})^2} + C$

26. $\arcsin(x - 1) + C$

27. $\dfrac{1}{2}\ln\left|\dfrac{e^x - 1}{e^x + 1}\right| + e^{-x} + C$

28. $\ln\dfrac{\sqrt{e^x + 1} - 1}{\sqrt{e^x + 1} + 1} + C$

29. $-\dfrac{\operatorname{sgn} x}{3}\left(1 + \dfrac{2}{x}\right)^{3/2} + C$

30. $\dfrac{2}{3}\arctan\sqrt{x^3 - 1} + C$

Appendix A

Table of Basic Integrals

1. $\int 0 \, dx = C$ on $(-\infty, \infty)$.

2. $\int x^n \, dx = \dfrac{x^{n+1}}{n+1} + C \ (n \neq -1)$

   on interval(s) depending on the exponent n.

3. $\int \dfrac{1}{x} \, dx = \ln |x| + C$ on each of the intervals $(-\infty, 0)$, $(0, \infty)$.

4. $\int a^x \, dx = \dfrac{a^x}{\ln a} + C \ (a > 0, a \neq 1)$ on $(-\infty, \infty)$.

   In particular, $\int e^x \, dx = e^x + C$ on $(-\infty, \infty)$.

5. $\int \sin x \, dx = -\cos x + C$ on $(-\infty, \infty)$.

6. $\int \cos x \, dx = \sin x + C$ on $(-\infty, \infty)$.

7. $\int \sec^2 x \, dx = \tan x + C$ on each of the intervals

   $(-\pi/2 + n\pi, \pi/2 + n\pi), \ n \in \mathbb{Z} := \{0, \pm 1, \pm 2, \ldots\}$.

8. $\int \csc^2 x \, dx = -\cot x + C$ on each of the intervals $(n\pi, \pi + n\pi), \ n \in \mathbb{Z}$.

9. $\int \sec x \tan x \, dx = \sec x + C$ on the same intervals as in 7.

10. $\int \csc x \cot x \, dx = -\csc x + C$ on the same intervals as in 8.

11. $\int \tan x \, dx = \ln |\sec x| + C = -\ln |\cos x| + C$

on the same intervals as in 7.

12. $\int \cot x \, dx = \ln|\sin x| + C$ on the same intervals as in 8.

13. $\int \sec x \, dx = \ln|\sec x + \tan x| + C$ on the same intervals as in 7.

14. $\int \csc x \, dx = -\ln|\csc x + \cot x| + C = \ln|\csc x - \cot x| + C$

on the same intervals as in 8.

15. $\int \sinh x \, dx = \cosh x + C$ on $(-\infty, \infty)$.

16. $\int \cosh x \, dx = \sinh x + C$ on $(-\infty, \infty)$.

For $a > 0$,

17. $\int \dfrac{1}{x^2 + a^2} \, dx = \dfrac{1}{a} \arctan \dfrac{x}{a} + C$ on $(-\infty, \infty)$.

18. (*"tall logarithm"*) $\int \dfrac{1}{x^2 - a^2} \, dx = \dfrac{1}{2a} \ln\left|\dfrac{x - a}{x + a}\right| + C$

on each of the intervals $(-\infty, -a)$, $(-a, a)$, (a, ∞).

19. $\int \dfrac{1}{\sqrt{a^2 - x^2}} \, dx = \arcsin \dfrac{x}{a} + C$ on $(-a, a)$.

20. (*"long logarithm"*) $\int \dfrac{1}{\sqrt{x^2 \pm a^2}} \, dx = \ln|x + \sqrt{x^2 \pm a^2}| + C$

on $(-\infty, \infty)$ for "+"

and on each of the intervals $(-\infty, -a)$, (a, ∞) for "−".

21. $\int \dfrac{1}{x\sqrt{x^2 - a^2}} \, dx = \dfrac{1}{a} \operatorname{arcsec}\left|\dfrac{x}{a}\right| + C$

on each of the intervals $(-\infty, -a)$, (a, ∞).

Appendix B

Reduction Formulas

For $n = 2, 3, \ldots$,

1. $\displaystyle \int \cos^n x \, dx = \frac{\cos^{n-1} x \sin x}{n} + \frac{n-1}{n} \int \cos^{n-2} x \, dx$

2. $\displaystyle \int \sin^n x \, dx = -\frac{\sin^{n-1} x \cos x}{n} + \frac{n-1}{n} \int \sin^{n-2} x \, dx$

3. $\displaystyle \int \sec^n x \, dx = \frac{\sec^{n-2} x \tan x}{n-1} + \frac{n-2}{n-1} \int \sec^{n-2} x \, dx$

4. $\displaystyle \int \csc^n x \, dx = -\frac{\csc^{n-2} x \cot x}{n-1} + \frac{n-2}{n-1} \int \csc^{n-2} x \, dx$

5. $\displaystyle \int \tan^n x \, dx = \frac{\tan^{n-1} x}{n-1} - \int \tan^{n-2} x \, dx$

6. $\displaystyle \int \cot^n x \, dx = -\frac{\cot^{n-1} x}{n-1} - \int \cot^{n-2} x \, dx$

7. For $n = 1, 2, \ldots$, $\displaystyle \int \ln^n x \, dx = x \ln^n x - n \int \ln^{n-1} x \, dx$

For $a \neq 0$ and $n = 1, 2, \ldots$,

8. $\displaystyle \int x^n e^{ax} \, dx = \frac{x^n e^{ax}}{a} - \frac{n}{a} \int x^{n-1} e^{ax} \, dx$

9. $\displaystyle \int x^n \cos ax \, dx = \frac{x^n \sin ax}{a} - \frac{n}{a} \int x^{n-1} \sin ax \, dx$

10. $\displaystyle \int x^n \sin ax \, dx = -\frac{x^n \cos ax}{a} + \frac{n}{a} \int x^{n-1} \cos ax \, dx$

For $a > 0$ and $n = 2, 3, \ldots,$

11. $\displaystyle\int \frac{1}{(x^2 + a^2)^n}\,dx$

$$= \frac{1}{(2n-2)a^2}\left[\frac{x}{(x^2+a^2)^{n-1}} + (2n-3)\int \frac{1}{(x^2+a^2)^{n-1}}\,dx\right]$$

Appendix C

Basic Identities of Algebra and Trigonometry

Algebra

Binomial Formula

For $n = 1, 2, 3, \ldots,$

$$(A \pm B)^n = \sum_{k=0}^{n} \binom{n}{k} A^{n-k} B^k = A^n + \binom{n}{1} A^{n-1} B + \cdots + \binom{n}{n-1} AB^{n-1} + B^n,$$

where $\binom{n}{k} := \dfrac{n!}{k!(n-k)!}$, $k = 0, 1, \ldots, n.$

In particular,

$$(A \pm B)^2 = A^2 \pm 2AB + B^2 \qquad \textit{(Perfect Square)}$$
$$(A \pm B)^3 = A^3 \pm 3A^2B + 3AB^2 \pm B^3 \qquad \textit{(Perfect Cube)}$$

Factoring Formula

For $n = 1, 2, 3, \ldots,$

$$A^n - B^n = (A - B) \sum_{k=0}^{n-1} A^{n-1-k} B^k$$
$$= (A - B)(A^{n-1} + A^{n-2}B + \cdots + AB^{n-2} + B^{n-1}).$$

In particular,

$$A^2 - B^2 = (A - B)(A + B) \qquad \textit{(Difference of Squares)}$$
$$A^3 \pm B^3 = (A \pm B)(A^2 \mp AB + B^2) \qquad \textit{(Sum/Difference of Cubes)}$$

Laws of Exponents

1. $A^m A^n = A^{m+n}$ (*Product Rule*)

2. $\dfrac{A^m}{A^n} = A^{m-n}$ (*Quotient Rule*)

3. $(A^m)^n = A^{mn}$ (*Power Rule*)

4. $(AB)^n = A^n B^n$ (*Power of Product Rule*)

5. $\left(\dfrac{A}{B}\right)^n = \dfrac{A^n}{B^n}$ (*Power of Quotient Rule*)

Laws of Logarithms

1. $\log_a(AB) = \log_a A + \log_a B$ (*Product Rule*)

2. $\log_a \dfrac{A}{B} = \log_a A - \log_a B$ (*Quotient Rule*)

3. $\log_a A^c = c \log_a A$ (*Power Rule*)

Change of Base Formula

$$\log_b A = \frac{\log_a A}{\log_a b}$$

In particular, $\log_b a = \dfrac{1}{\log_a b}$.

Trigonometry

Reciprocal Identities

$$\sin\theta = \frac{1}{\csc\theta} \qquad \cos\theta = \frac{1}{\sec\theta} \qquad \tan\theta = \frac{1}{\cot\theta}$$

$$\csc\theta = \frac{1}{\sin\theta} \qquad \sec\theta = \frac{1}{\cos\theta} \qquad \cot\theta = \frac{1}{\tan\theta}$$

Quotient Identities

$$\tan \theta = \frac{\sin \theta}{\cos \theta} \qquad \cot \theta = \frac{\cos \theta}{\sin \theta}$$

Pythagorean Identities

$$\sin^2 \theta + \cos^2 \theta = 1 \qquad \tan^2 \theta + 1 = \sec^2 \theta \qquad \cot^2 \theta + 1 = \csc^2 \theta$$

Even-Odd Identities

$$\sin(-\theta) = -\sin \theta \qquad \cos(-\theta) = \cos \theta \qquad \tan(-\theta) = -\tan \theta$$

$$\csc(-\theta) = -\csc \theta \qquad \sec(-\theta) = \sec \theta \qquad \cot(-\theta) = -\cot \theta$$

Double-Angle Identities

$$\sin 2\theta = 2 \sin \theta \cos \theta$$
$$\cos 2\theta = \cos^2 \theta - \sin^2 \theta = 2\cos^2 \theta - 1 = 1 - 2\sin^2 \theta$$

Power Reduction/Half-Angle Identities

$$\sin^2 \theta = \frac{1 - \cos 2\theta}{2} \qquad \cos^2 \theta = \frac{1 + \cos 2\theta}{2} \qquad \tan^2 \theta = \frac{1 - \cos 2\theta}{1 + \cos 2\theta}$$

Product-to-Sum Identities

$$\cos \alpha \cos \beta = \frac{1}{2}[\cos(\alpha - \beta) + \cos(\alpha + \beta)]$$

$$\sin \alpha \sin \beta = \frac{1}{2}[\cos(\alpha - \beta) - \cos(\alpha + \beta)]$$

$$\sin \alpha \cos \beta = \frac{1}{2}[\sin(\alpha - \beta) + \sin(\alpha + \beta)]$$

Bibliography

[1] W. Briggs, L. Cochran, B. Gillett, et al., *Calculus: Early Transcendentals*, 2nd ed., Pearson Education, Inc., Boston, 2015.

[2] V.F. Butuzov, N.Ch. Krutitskaya, G.N. Medvedev, and A.A. Shishkin, *Mathematical Analysis in Questions and Problems*, 4th ed., Fizmatlit, Moscow, 2001 (Russian).

[3] B.P. Demidovich, *Collection of Problems and Exercises in Mathematical Analysis*, 13th ed., Publishing House of Moscow University, Moscow, 1997 (Russian).

[4] B.R. Gelbaum and J.M.H. Olmsted, *Counterexamples in Analysis*, Dover Publications, Inc., Mineola, New York, 2003.

[5] V.P. Minorsky, *Collection of Problems in Higher Mathematics*, Fizmatlit, Moscow, 2006 (Russian).

[6] J. Stewart, *Calculus*, 3rd ed., Brooks/Cole Publishing Co., Pacific Grove, California, 1995.

Index

Printed in the United States
By Bookmasters